1796
J61
T5
1965
v.3

ATLAS OF CEREAL DISEASES AND PESTS IN EUROPE

Agro-ecological Atlas of Cereal Growing in Europe

Volume I
AGRO-CLIMATIC ATLAS OF EUROPE

Volume II
ATLAS OF THE CEREAL-GROWING AREAS IN EUROPE

Volume III
ATLAS OF CEREAL DISEASES AND PESTS IN EUROPE

Agro-ecological Atlas of Cereal Growing in Europe

Volume III

ATLAS OF CEREAL DISEASES AND PESTS IN EUROPE

J.C. ZADOKS and F.H. RIJSDIJK

Department of Phytopathology
Agricultural University, Wageningen, Netherlands

Pudoc Wageningen 1984

The Agro-ecological Atlas of Cereal Growing in Europe has 3 volumes:
Volume I, Agro-climatic Atlas of Europe (ISBN 0-444-40569-0) and
Volume II, Atlas of the Cereal-Growing Areas in Europe (ISBN 0-444-40819-3),
published jointly by Pudoc, Wageningen and Elsevier, Amsterdam, are distributed by Elsevier Scientific Publishing Company, Amsterdam, Oxford, New York.
Volume III, Atlas of Cereal Diseases and Pests in Europe, is published and distributed by Pudoc, Wageningen.

CIP

Zadoks, J.C.

Atlas of cereal diseases and pests in Europe / J.C. Zadoks
and F.H. Rijsdijk. – Wageningen: Pudoc. – Ill., krt. –
(Agro-ecological atlas of cereal growing in Europe; vol. 3)

Met lit. opg., index.
ISBN 90-220-0863-0 geb.
SISO 632.6 UDC 632:633.1 (4)
Trefw.: planteziekten; landbouw; Europa.

ISBN 90 220 0863 0

© Centre for Agricultural Publishing and Documentation (Pudoc), Wageningen, 1984

No part of this publication, apart from bibliographic data and brief quotations embodied in critical reviews, may be reproduced, re-recorded or published in any form including print, photocopy, microfilm, electronic or electromagnetic record without written permission from the publisher Pudoc, P.O. Box 4, 6700 AA Wageningen, Netherlands.

Printed in the Netherlands

Contents

European Cereal Atlas Foundation	6
In memoriam Dr Ir W. Feekes	7
In memoriam Dr S. Broekhuizen	8
Preface	9
Acknowledgments	10
List of abbreviations	11
Mapping of losses	12
Sources of information	12
International survey	13
Review of the literature	14
Personal observations	15
Handling of data	16
Geographic units	16
Weighting the data	16
From damage code to proportion of damage	17
Mapping procedure	18
The maps	19
Value to be attached to damage maps	20
References	21
Maps	23
Index of scientific names	165
Index of common names	167

European Cereal Atlas Foundation

The European Cereal Atlas Foundation, ECAF, is a non-profit organization aiming at the promotion of scientific research on cereals and their grains, firstly by compiling an Atlas of Cereal Growing in Europe, and further by making any other publications about cereals and by stimulating scientific work on cereals which anticipate the creation of publications.

ECAF Board

Dr Ir C. MASTENBROEK	Chairman (Dronten, Netherlands)
Dr Ir W. LANGE	Secretary (Foundation for Agricultural Plant Breeding, Wageningen, Netherlands)
Mr A. RUTGERS	Treasurer (Ministry of Agriculture and Fisheries, Netherlands)
Dr F.G.H. LUPTON	Member (Plant Breeding Institute, Cambridge, England)
Ms Ir A. PRINS	Member (Netherlands Grain Centre, Wageningen, Netherlands)
Dr A. BRÖNNIMANN	Member (Swiss Federal Research Station for Agronomy, Zürich-Reckenholz, Switzerland)

In memoriam Dr Ir W. Feekes

27 December 1907 – 9 February 1979

Dr Ir W. Feekes was an agricultural scientist. He gained his doctorate in botany, a significant fact in his career. He spent most of his life in studying wheat. His originality and vision were based on a deep respect for all living beings. He was a man of grand ideas and broad approaches. For him, the Netherlands was too small a working area: he was a European, travelling all over Europe every year, meeting business relations and friends at the same time. As co-founder and first President of the Netherlands Grain Centre, he contributed much to research on cereals in and outside the Netherlands. One of his comprehensive ideas was to publish an Agro-ecological Atlas of Cereal Growing in Europe, of which this is Volume III. For technical organization and financing of the Atlas, Dr Feekes established the European Cereal Atlas Foundation, of which he was the first Chairman. Dr Feekes' enthusiasm was contagious. He always found people willing to carry through his ideas. Once the project under way, he followed it with keen interest. His stimulatory remarks kept people working, because he had a gift to combine deep scientific understanding with practical applicability. He was much concerned about the delays in the preparation of the present Volume, but he lived to see the work resumed. The authors are greatly indebted to Dr Feekes for training and inspiration and regret that he is no witness to the birth of his brain child, the Atlas of Cereal Diseases and Pests in Europe.

In memoriam Dr S. Broekhuizen

8 December 1903 – 10 March 1971

Dr S. Broekhuizen began work as a biologist trained in plant pathology. From 1950 until his retirement in 1969, he was associated with research on cereals, lastly as the Expert-Secretary of the Netherlands Grain Centre. In this capacity, he established an international network of knowledge, trust and friendship among cereal scientists all over Europe. His expertise was widely acknowledged and much in demand. He collected ideas, gave them flesh and bones by shaping them into projects, worked his way through the red tape, at times so well that he wove it into a red carpet. He translated his enthusiasm into the perseverance needed to get jobs done. Being extremely efficacious in his work, Dr Broekhuizen saw tasks through to the end. Besides his many administrative duties, he took an active interest in the research going on. After retirement, he simply worked on. Within a year, he had published the Agro-ecological Atlas of Cereal Growing in Europe, Volume II 'Atlas of the Cereal-growing Areas in Europe'. The present Volume was first conceived in 1968, with the assistance from the present senior author. Dr Broekhuizen was delighted to return to plant pathology at the end of his career. The junior author assisted him during his last illness. After his death, the authors continued the work with many interruptions, due to changes of occupation and to an increasingly unfavourable administrative climate, urged on by the desire to do justice to Dr Broekhuizen's work.

Preface

'Since it is expected that closer economic collaboration between various European states will also lead to a freer exchange of the varieties of different agricultural crops, the need has arisen among breeders for more up-to-date information on soil and climatic conditions in the agricultural areas of Europe and on the pattern of development of the crops in these areas. It is not only the breeders who are so concerned, but also seed merchants, variety testing authorities and various advisory services'.

These opening words of Volume I apply today even more than in 1965, when they were written. The idea of producing a series of Agro-ecological Atlases of Cereal Growing in Europe ripened in the nineteen fifties. It first took form with the installation of a 'Climate Atlas Study Group' in 1961, by the European Association for Research on Plant Breeding (Eucarpia).

The first Volume was published in 1965; the second in 1969. The prefaces to these atlases clearly show how ideas on the design of the atlases changed gradually. The second atlas and the present one have been designed and produced under the aegis of the European Cereal Atlas Foundation (ECAF), an independent non-profit inter-European trust established in 1968.

Though the opening sentence of Volume I quoted above is still true, and chemical industry and public health authorities can be added to the list of interested parties, much has changed. The economic situation of Europe is less optimistic than fifteen years ago, as was painfully felt during the preparation and production of this volume. The Volume was prepared with a bare minimum of money, and most of the work was carried out in spare time. The much regretted death of Dr S. Broekhuizen in 1971 and the transfer of the responsibility to the present authors contributed to the delay in publishing of this Volume, which had been initiated in 1968. Economic considerations enforced a smaller size and a simpler design, with texts in English only.

Thanks to the great optimism and continuous hopeful support of the Board of ECAF, and thanks to the unselfish collaboration of many people, the authors were encouraged to persevere in their efforts. The data used mainly come from the decade 1961-1970. Though the data are well seasoned by now, the results are still useful and applicable today. There seem to have been few changes on a European scale over the decade 1971-1980. Noteworthy exceptions are damage from cereal aphids and barley net blotch.

Wageningen, 1984
J.C. Zadoks
F.H. Rijsdijk

Acknowledgments

Countless colleagues have contributed to this atlas. It makes no sense to pick out any for special mention. Suffice it to mention two broad categories: the staff in Wageningen and the informants all over Europe. In addition, there are the initiators, the late Dr W. Feekes and the late Dr S. Broekhuizen, the board members of ECAF and, at an early stage, the members of the Advisory Committee. Mr W.C.Th. Middelplaats patiently and very carefully drafted all maps into their present form.

The publication of this Atlas was made possible by generous grants from:
- Netherlands Grain Centre (NGC), Wageningen
- Pudoc, Wageningen
- Agricultural University, Wageningen
- Agricultural University Special Fund, Wageningen
- Granaria BV, Rotterdam
- BASF-Nederland BV, Arnhem
- Fund for the Promotion of Breeding Field Crops, Wageningen
- European Cereal Atlas Foundation (ECAF), Wageningen

List of abbreviations

AFREQ	Average FREQuency per GEON
BUIN	Basic Unit of INformation
FREQ	FREQuency of occurrence per BUIN
DACO	DAmage COde per BUIN
ECAF	European Cereal Atlas Foundation, Wageningen
GEON	GEOgraphic uNit
ISO	ISOlines of damage
MAD	Mean Annual Damage (usually expressed as 1000 x MAD)
PDAM	Proportional DAMage
WADACO	Weighted Average of DACO
WAPDAM	Weighted Average PDAM

Mapping of losses

Crop loss maps are thematic maps. They present information on a theme, here crop loss, against a geographic background. A vast amount of diverse information can be condensed and presented in maps in easily accessible form. The handling of the information from input to output must follow certain rules of procedure, which will be explained here in brief.

First, the various sources of information will be explained; second, data handling will be discussed; and finally some comments will be given on the value of the results. Figure 1 is a simple block diagram of procedure.

Sources of information

The input information came from three sources:
- international survey by questionnaire
- review of the literature
- personal observations.

Before describing these sources in detail, a word has to be said on their quality. Most of the input information was 'soft', which means that it was based on personal impressions, with vague indications of area, time, severity, or amount of damage. Quantitative information was rare. In fact, most of the figures received had to be read as short-hand indications of rather broad generalizations.

Information of the type produced by the Harpenden School (James, 1974; Large, 1966) is regarded as hard information. Such information was rare for Europe as a

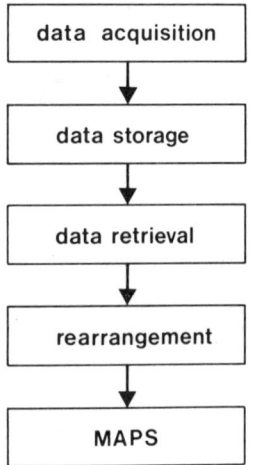

Fig. 1. Simple block diagram of data bank on crop losses. From Rijsdijk & Zadoks, 1979.

whole. Nevertheless, a closer inspection of the available data revealed a range from 'typically hard' to 'very soft'.

Any ranking of information on a 'hardness' scale is highly subjective. Nevertheless, rules of procedure were sought to attain ranking. The outcome must be to attach a weighting to every single unit of information. The weightings are themselves subjective. However the method has the merit of inviting debate. The outcome of such a debate could be that procedures are modified and weightings adjusted. So the procedure is described in detail.

The authors regret that some major sources of hard information were not disclosed to them:
– records of field trials held by the pesticide industry and by the pesticide registration offices
– records of field trials by institutes testing new cultivars.

International survey

For the international survey on diseases and pests of cereals in Europe, a questionnaire was designed and distributed in 1972. The forms were distributed among colleagues working in research institutes, breeding stations, extension services, and so on, all over Europe. A total of 192 forms was returned, with one to ninety lines completed.

The information requested consisted of three parts:
(a) identification
(b) 10-year data
(c) single-year data.
A fourth part was added by the authors:
(d) weighting factors.

A basic unit of information (BUIN) consisted of one 'line' containing the identification Section a, including the pathosystem recorded, Sections b and c, which were usually incomplete, and the weighting factors of Section d. Blanks were considered as absence of information, not as absence of the harmful agent. Every single BUIN was copied onto a punched card, some 22 000 cards in total.

The identification section of a BUIN consists of nine items (Table 1). For the longitude east, the minus sign was omitted. The data referred either to a 10-year period, and Column 6 was assigned Code 1, or to a particular year specified in Column 9 by its last two digits, with Code 2 in Column 6.

The 10-year data consisted of:
– frequency of occurrence
– mean annual damage
– highest damage observed
– increase or decrease during the period.

For each harmful agent, the frequency of occurrence had been asked for. The frequency (FREQ) was classed into four groups:
1. rarely seen
3. not infrequent
5. rather frequent
7. nearly always seen.

Intermediate values were acceptable. Obviously, the frequency data depend on the knowledge and accuracy of the informant. The differences between informants were

Table 1

Identification section of a basic unit of information (BUIN)

Item	Digits	Comment
1 Country	2	
2 Informant	3	The scientist or group of scientists completing the form; publication number.
3 Latitude N	2+2	Degrees and minutes.
4 Longitude E	2+2	Degrees and minutes.
5 Climatic subprovince		Added from Volume I of the Agro-ecological Atlas of Cereal Growing in Europe.
6 Period	1	
7 Crop	1	
8 Harmful agent	3	
9 Year	2	Only for single-year data.

great. No standardization was attempted.

The mean annual damage was characterized by a damage code (DACO):
1. light
3. moderate
5. severe.

Checks could be made by comparing neighbouring informants.

The greatest damage observed was recorded in per cent. In the record, if any, the informant's personal knowledge (and emotions!) are involved. The reference yield is not known.

The 10-year data requested included indication of increase/decrease in the harmful agent. The answers received were so incomplete that they could not be used.

Single-year data included mention of specific years with conspicuous attacks. The severity of the damage was to be recorded in the three classes mentioned above. The information received was used for two purposes:
− to calculate a relation between damage and severity
− to interpret frequency data.

Review of the literature

The literature was reviewed by Pudoc. This review covered many papers from nearly all European countries, often written in the national language, over the period 1946-1970, in which crop losses were mentioned. The information thus obtained is less systematic and less uniform than the information obtained from the survey. The information is subject to far more serious doubts than those on the survey. Nevertheless, the literature can be analysed into basic units of information, coded as described above and punched onto cards, one card per pathosystem. Most published data were for single years.

Personal observations

Personal observations by the authors and some advisors were used mainly as a check, rarely as input.

Handling of data

Geographic units

For mapping, all information was sorted by geographic unit. The geographic unit (GEON) chosen is the climatic subprovince as defined in Volumes I and II (Thran & Broekhuizen, 1965; Broekhuizen, 1969), Map 1. These subprovinces were delimited on the basis of climatic data. Within each subprovince, the climate is reasonably uniform with respect to the ecology of harmful agents.

The number of BUINs for several combinations of pathosystem and subprovince was low, so that subprovinces had to be combined into larger geographic units. For example, the 82 subprovinces of the European Soviet Union were combined into seven GEONs. Many geographic units remained empty due to lack of information.

Weighting the data

Compaction of the vast amount of input data into relatively simple maps with damage isolines necessitated standardized computational procedures. The diversity of the input data was such that a classification had to be made in accordance with their relevance for mapping of damage. Weightings had to be assigned to all classes thus distinguished.

A geographical criterion for classification was the size of the area to which the data applied. This size needed to be related to the size of the geographic units, which is somewhere between that of a small country and that of an intermediate administrative unit. After due consideration, the authors applied the following weighting factors:
2: local data (part of field, field, or immediate environment of informant)
6: data from intermediate administrative levels
3: federal and national data.

A temporal criterion was the age of the data, according to source year. Weighting by age was necessary to balance the data from the literature survey (1961-1970).

Old data were not useless but were less relevant. The weighting factors were
1:      source year $<$ 1951
2: 1950 $<$ source year $<$ 1966
6: 1965 $<$ source year.

The source year was, as far as possible, the year to which the data referred, or, if not stated, the year of publication.

The data processing as described in this and the following section is illustrated in Figure 2. The weighting procedure is summarized in Table 2.

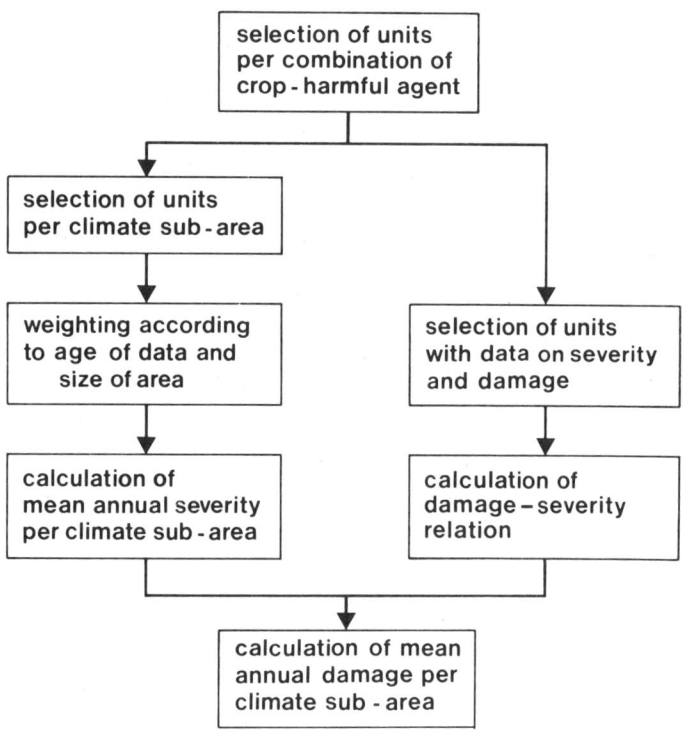

Fig. 2. Simplified block diagram of data handling including weighting procedure. From Rijsdijk & Zadoks, 1979.

From damage code to proportion of damage

Some informants quoted damage in per cent. Per cent of what? The authors assumed that the informants used 'attainable yield' as a reference. Translating per cent as proportion, this figure is called proportional damage (PDAM). For many BUINs, PDAM is known.

Most informants gave damage codes. Damage codes and proportional damage could be related by certain rules:
– Select a pathosystem.
– Select those GEONs with at least one DACO and one PDAM. Select all BUINs with values of both DACO and PDAM > 0.
– Estimate the linear regression of PDAM on DACO and the regression coefficient.

Table 2

Weightings of basic units of information (BUIN) according to their geographical and temporal class.

Geographical class		Temporal class		
		< 1951	1950 < year < 1966	Decade 1961/1970 or 1965 < year
	Weight	1	2	6
Local	2	2	4	12
Intermediate	6	6	12	36
National	3	3	6	18

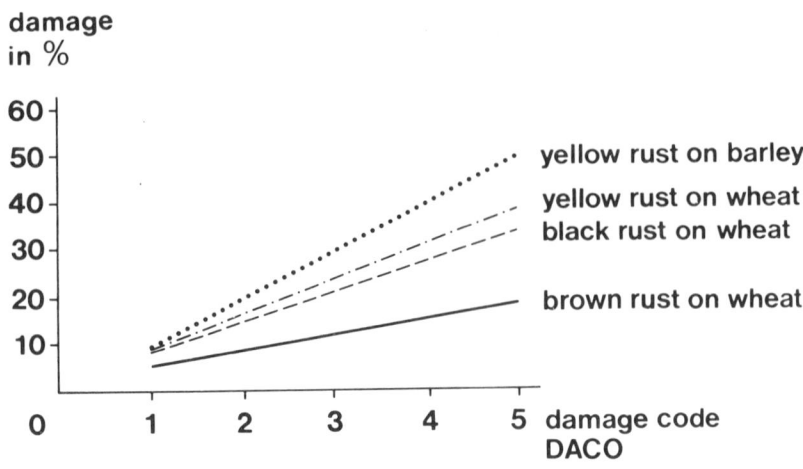

Fig. 3. Relation between damage code and damage calculated from the opinions of the informants. Abscissa — damage code: 1 = light; 3 = moderate; 5 = severe. Ordinate — damage (%). From Rijsdijk & Zadoks, 1976.

- Estimate the average value of DACO per GEON weighted for geographical and temporal criteria (WADACO).
- Estimate the average frequency per GEON (AFREQ) from the values FREQ per GEON and the single-year data per GEON.
- Estimate the mean annual damage per GEON (MAD) by multiplication of WADACO with AFREQ.

There were usually no low values of PDAM. To avoid steep regression lines due to this lack of data, the origin of the graph with the paired values (WAPDAM, WADACO) = (0,0) was always added to the set of data pairs before calculating the regression equation (Figure 3).

The procedure had no statistical implications; it was purely a computational device. The 'regression lines' were adjusted, when necessary, to go through the origin.

Mapping procedure

Several steps were needed to obtain computer-prepared lines of equal damage, or damage isolines:
- recording of GEON coordinates
- estimation of MAD for each combination of pathosystem and GEON
- calculation of guide points for isodams.

For each GEON, the coordinates were recorded and listed. This was a once-only operation. The coordinates chosen were those of the 'centre of gravity' of the GEON on the map. The centre of gravity was called the GEON centre.

For each combination of pathosystem and GEON MAD was calculated as described above and listed together with the GEON centre.

The mapping procedure went through the following steps:
- Select a pathosystem.
- Select a GEON with a MAD value.
- Find the five nearest GEONs with MAD values.

- For each of these five GEONs, find its distance in grid units (km) from the central GEON.
- For each of these five GEONs, calculate the difference in MAD from the central GEON.
- Along each of the lines connecting the five GEONs with the central GEON, calculate those guide points where the interpolated MAD values are multiples of 0.01.
- Record the coordinates of each of the guide points together with the MAD values.
- Transform the coordinates thus recorded by means of polar coordinates and correction factors so that the computer-coupled plotter can position the guide points on the map.
- Plot the GEON centre and the guide points on the map using for each point a symbol corresponding with its (interpolated) MAD value.
- Select the next GEON with a MAD value and go through the whole cycle again; continue to do so until all GEONs of the pathosystem have been dealt with.
- Plot the GEON centres, guide points, MAD values, and some geographic orientation points on an overlay to a standard geographic map.

The actual isolines were drawn by hand by connecting the guide points, with due consideration for the geographic and ecological characteristics of each area. The isolines thus constructed are called isodamage curves or shortly 'isodams'.

The maps

Over a hundred maps were prepared, 69 of which are actually published. The published maps fall in two categories, isodam maps and MAD maps. Maps containing insufficient information were rejected.

The original maps all gave mean annual damage (MAD). Only maps with MAD values for over 20 GEONs were accepted for publication. The reader should not be misled by the apparent accuracy of the MAD values. The first digit rather than the last one is significant, or, the order of magnitude is more relevant than the calculated value appearing in print.

Whenever possible, MAD maps were upgraded to isodam maps. Isodams connect points with equal MAD. Isodam maps were prepared only when the number of GEONs with MAD values was at least 30. Isodams could not always be drawn sensibly and, if so, maps are presented at MAD level. Intervals between isodams were regularly spaced between zero and the highest MAD value, but the spacing differs between maps. Most information is available from North-West and Central Europe. Several maps became vague towards the south or the east. Such scanty information as available was rendered by indications such as '0' or '<5', thus violating the isoline principle. The '>' and '<' signs were also used to complement the isodams and to avoid possible confusion.

To compare the two map representations an 'average' example was chosen, 'shield bugs', whose distribution was depicted as a MAD map and an isodam map (Maps 40 and 41).

Maps were arranged alphabetically by the taxonomic name of the harmful agent mapped. Two indexes, one for taxonomic and one for common English names, were included to ease access to the maps. Taxonomic names were based on recent publications in the following order of priority: Boerema & Verhoeven, 1977; Wiese, 1977; Buhl et al., 1975. For synonyms, the reader is referred to these publications. For common names, current synonyms are included at least in part.

Value to be attached to damage maps

The algorithms of the program were so designed that isodams appeared even with scanty information. The more information, the more reliable the damage map will be. But, the input data are mostly opinions, soft data, and only rarely they are experimental results, hard data. Maps based on opinions should be regarded as opinions too: opinions of the authors.

The authors used their experience to design weighting algorithms and to choose weighting parameters in the attempt to bring diverse data into a uniform frame. The algorithms and parameters, once chosen, were applied equally to all pathosystems, thereby establishing some degree of uniformity. By publishing the rules of procedure and the parameters here, they are open to scientific discussion. They can be changed, improved and applied again.

The authors feel confident that structured processing of opinions is possible and that it leads to valuable information that could not have been produced otherwise. A few maps were subjected to the opinion of specialists. Their approval, another subjective judgement, was encouraging. The procedures used have led to a degree of objectivity, that may be called objectivity by consensus (Rijsdijk & Zadoks, 1979).

The data-processing system ignored reference yield, as this was simply not known, or if known, its value in terms of this atlas is obscure. Uncertainty about yields was a handicap indeed, but not an insurmountable one. There is another handicap; damage is rarely caused by a single harmful agent, but interaction between harmful agents has been disregarded (Van der Wal et al., 1975).

Finally, the delay between data collection and map publication has been considerable. Several maps, including those for yellow rust of wheat and barley mildew, are still relevant. In other cases, such as wheat aphids and barley net blotch, the situation has changed considerably. The maps primarily reflect the period 1965-1970. The reader should judge subsequent changes for himself.

References

Boerema, G.H.; Verhoeven, A.A.: 1977. Check-list for scientific names of common parasitic fungi. Netherlands Journal of Plant Pathology 83: 165-204.

Broekhuizen, S. (Ed.): 1969. Agro-ecological atlas of cereal growing in Europe. II. Atlas of the cereal-growing areas in Europe. Pudoc, Wageningen, and Elsevier's Publ. Comp., Amsterdam, London, New York. 60 maps and 157 pp.

Buhl, C.; Weidner, H.; Zogg, H: 1975. Krankheiten und Schaedlinge an Getreide und Mais: ein Bestimmungsbuch. Eugen Ulmer, Stuttgart. 451 pp.

James, W.C.: 1974. Assessment of plant diseases and losses. Annual Review of Phytopathology 12: 27-48.

Large, E.C.: 1966. Measuring plant disease. Annual Review of Phytopathology 4: 9-28.

Rijsdijk, F.H.; Zadoks, J.C.: 1976. Assessment of risks and losses due to cereal rusts in Europe. Proceedings of the 4th European and Mediterranean Cereal Rust Conference, Interlaken: 60-62.

Rijsdijk, F.H.; Zadoks, J.C.: 1979. A data bank on crop losses: first experiences. EPPO (European and Mediterranean Plant Protection Organisation, Paris) Bulletin 9: 297-303.

Thran, D.; Broekhuizen, S.: 1965. Agro-ecological atlas of cereal growing in Europe. I. Agro-climatic atlas of Europe. Pudoc, Wageningen, and Elsevier's Publ. Comp., Amsterdam, London, New York. 125 maps and 38 pp.

Wal, A.F. van der; Smeitink, H.; Maan, G.C.: 1975. An ecophysiological approach to crop losses, examplified in the system wheat, rust, and glume blotch. III. Effects of soil-water potential on development, growth, transpiration, symptoms, and spore production of leaf rust-infected wheat. Netherlands Journal of Plant Pathology 81: 1-13.

Wiese, M.V.: 1977. Compendium of wheat diseases. American Phytopathological Society, St. Paul. 106 pp.

Maps

N means number of data. Figures on the maps represent mean annual damages per mill (MAD).
Isodams were drawn only when sufficient information was available.
Nearly all data were collected from the decade 1961-1970.

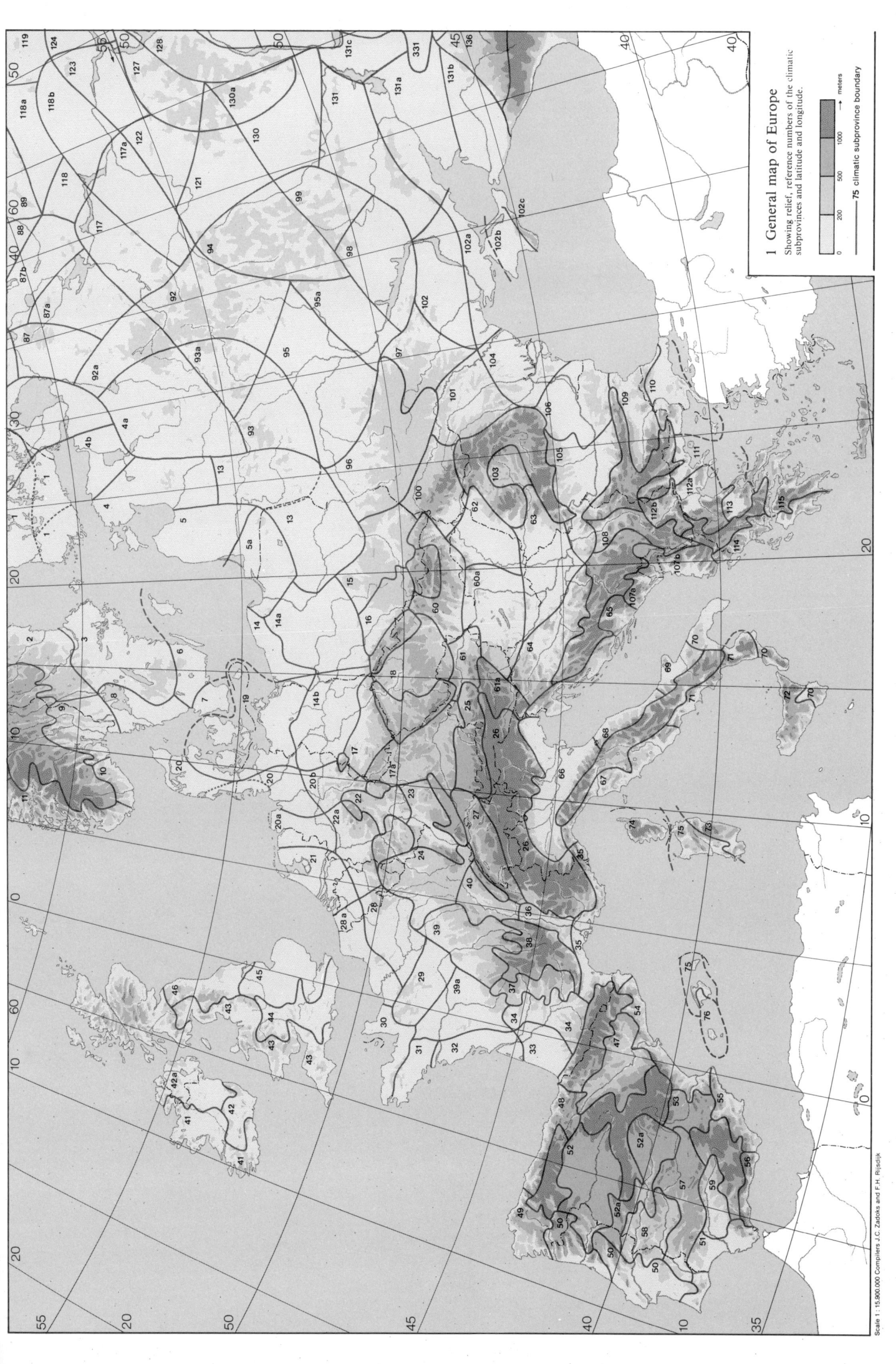

1 General map of Europe

Showing relief, reference numbers of the climatic subprovinces and latitude and longitude.

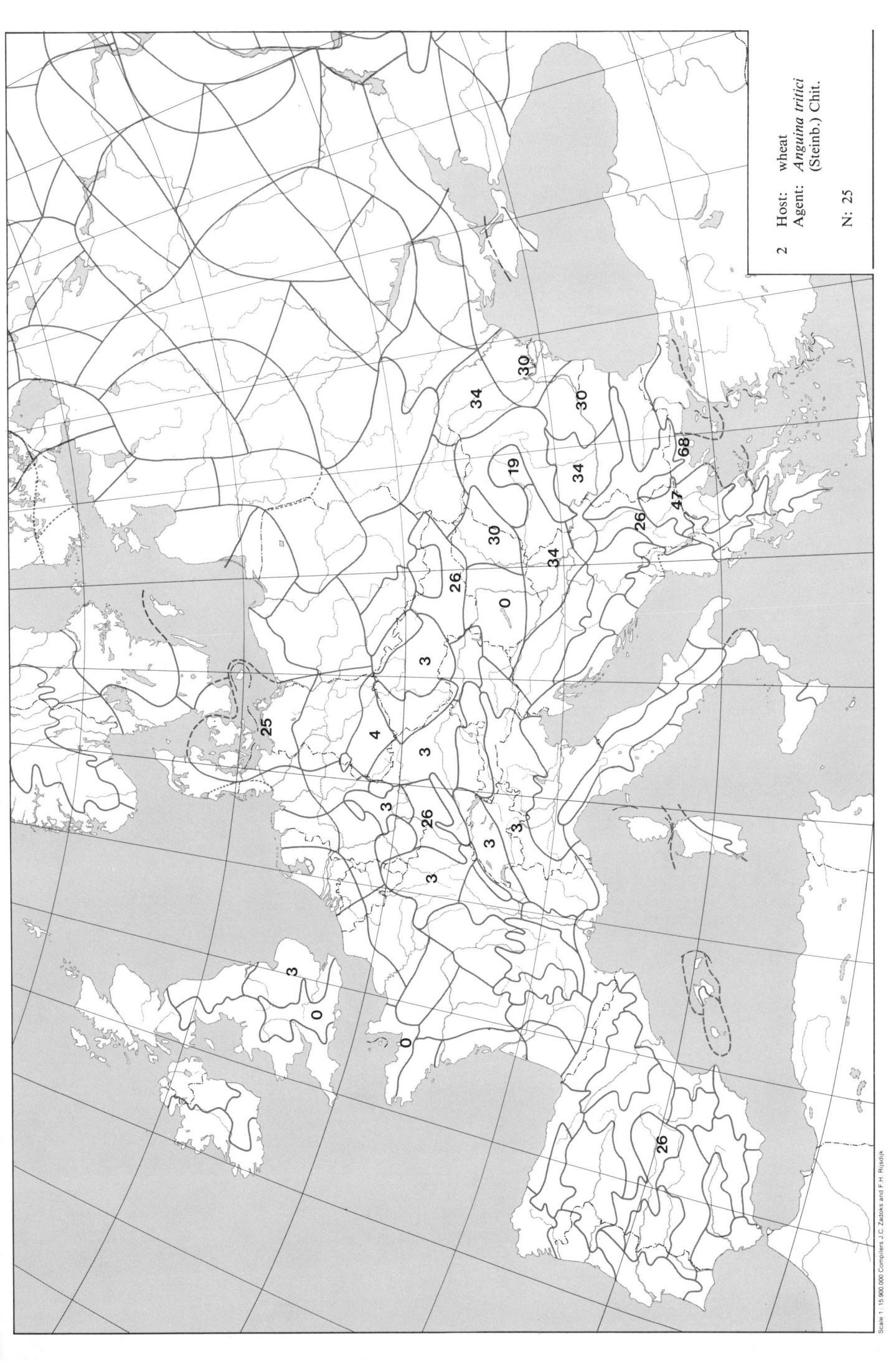

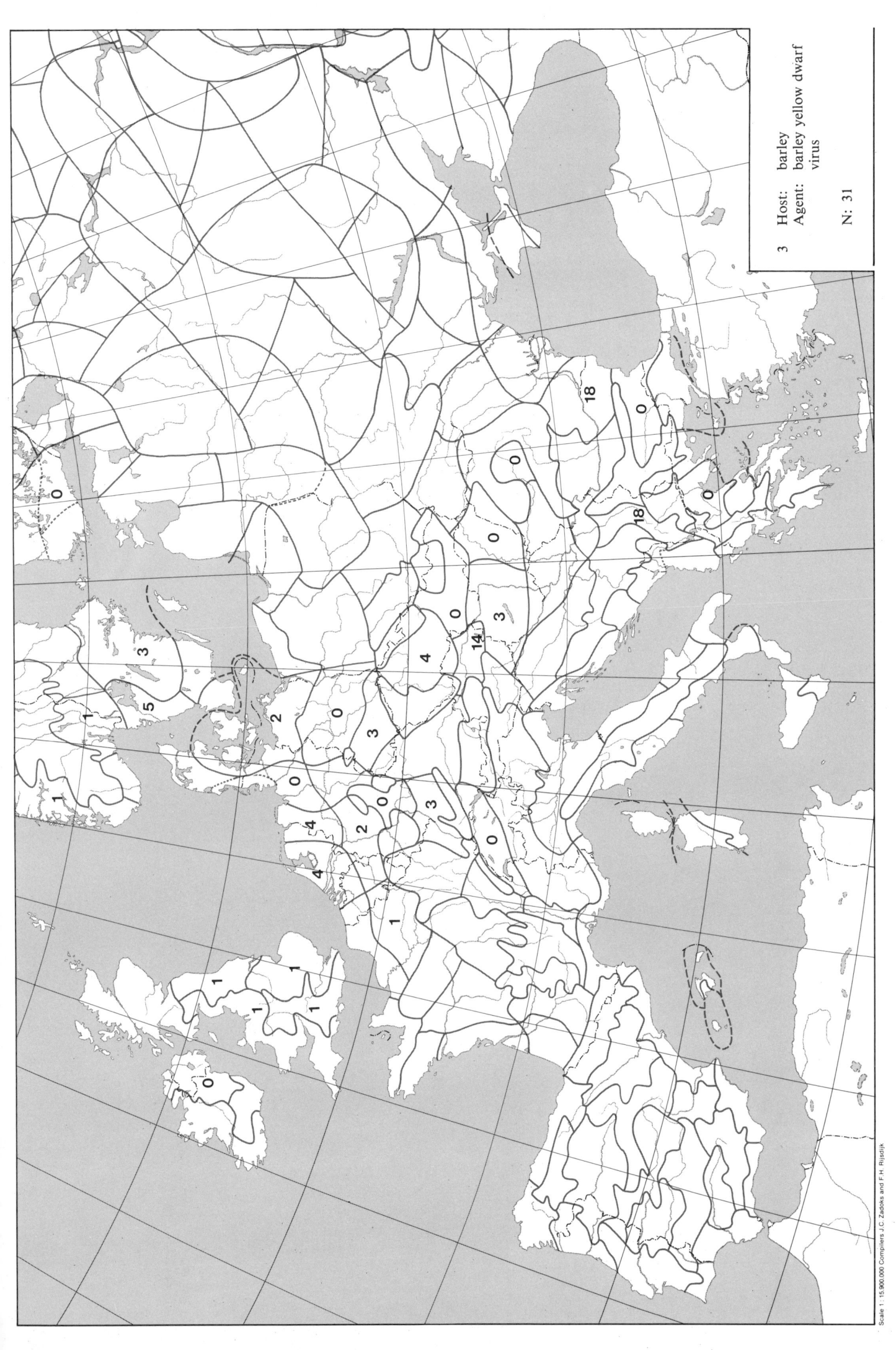

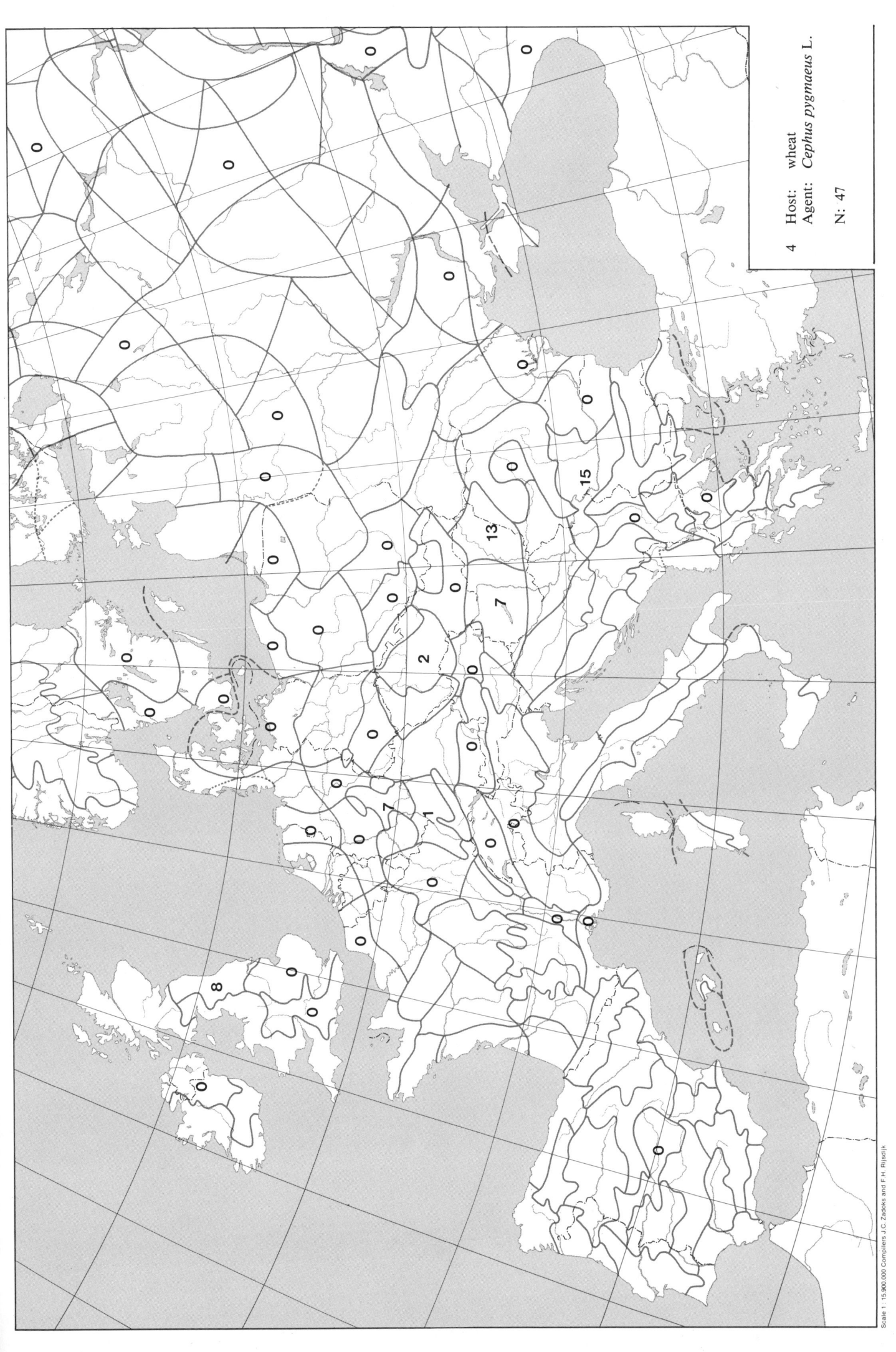

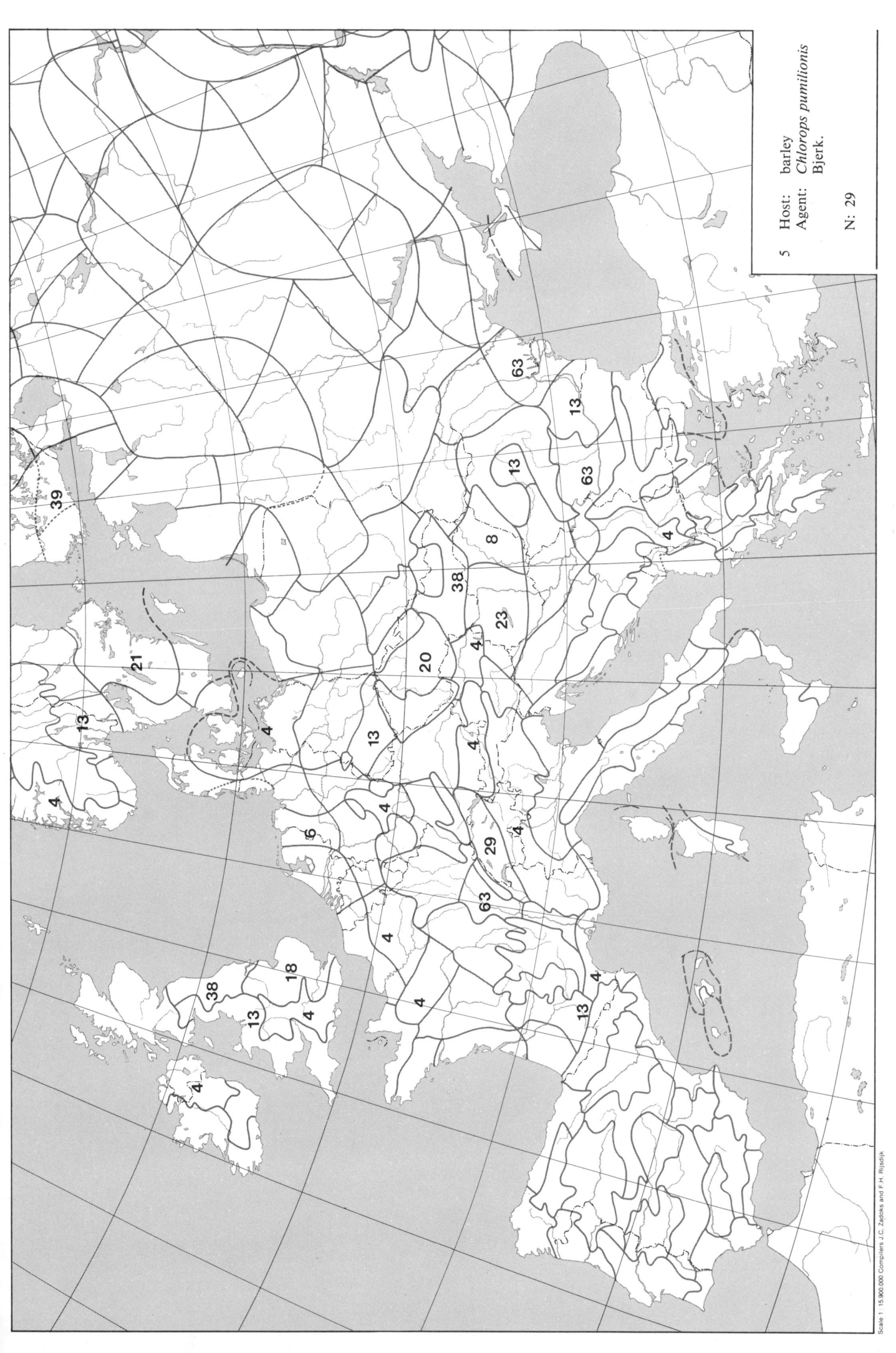

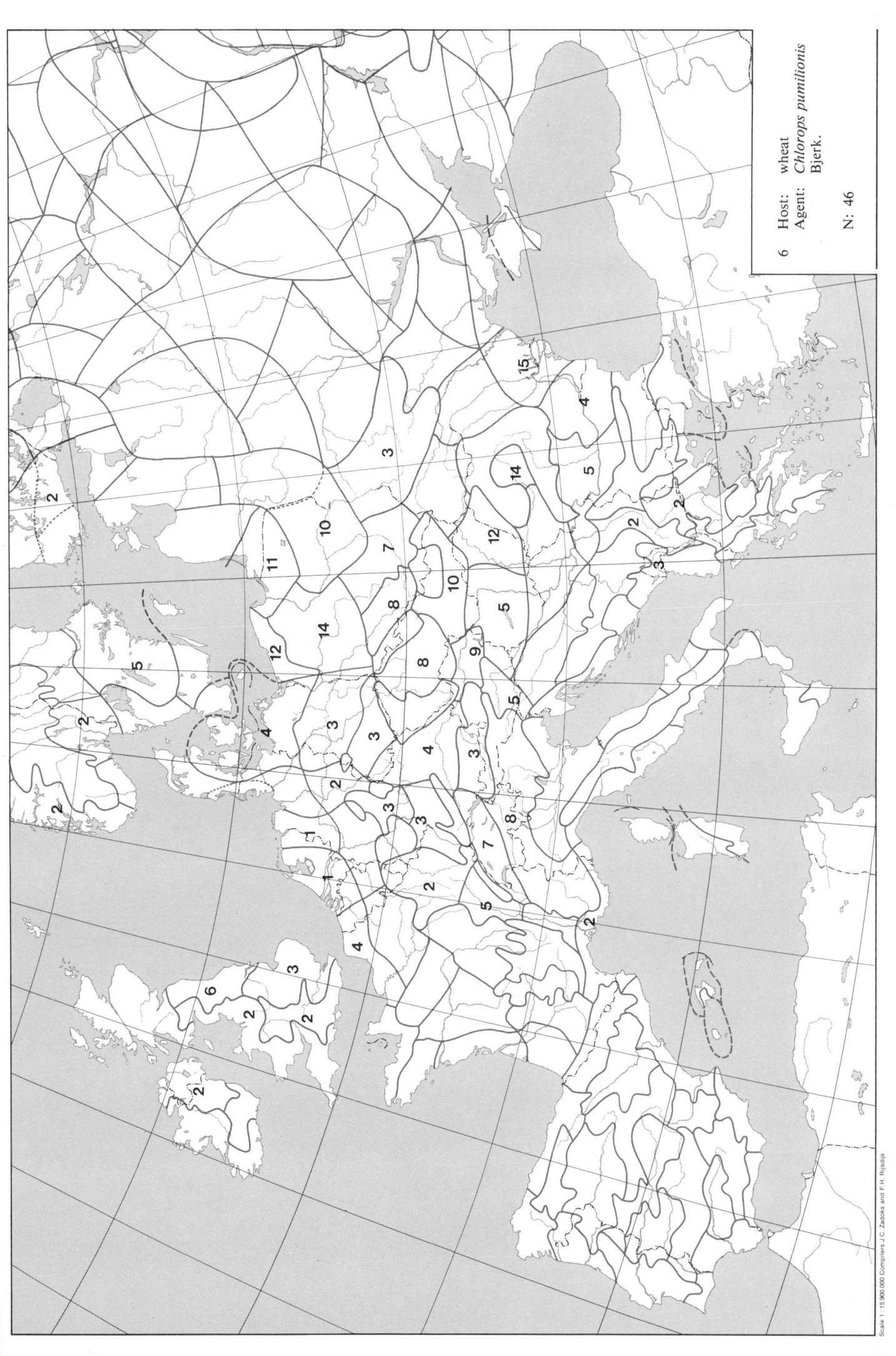

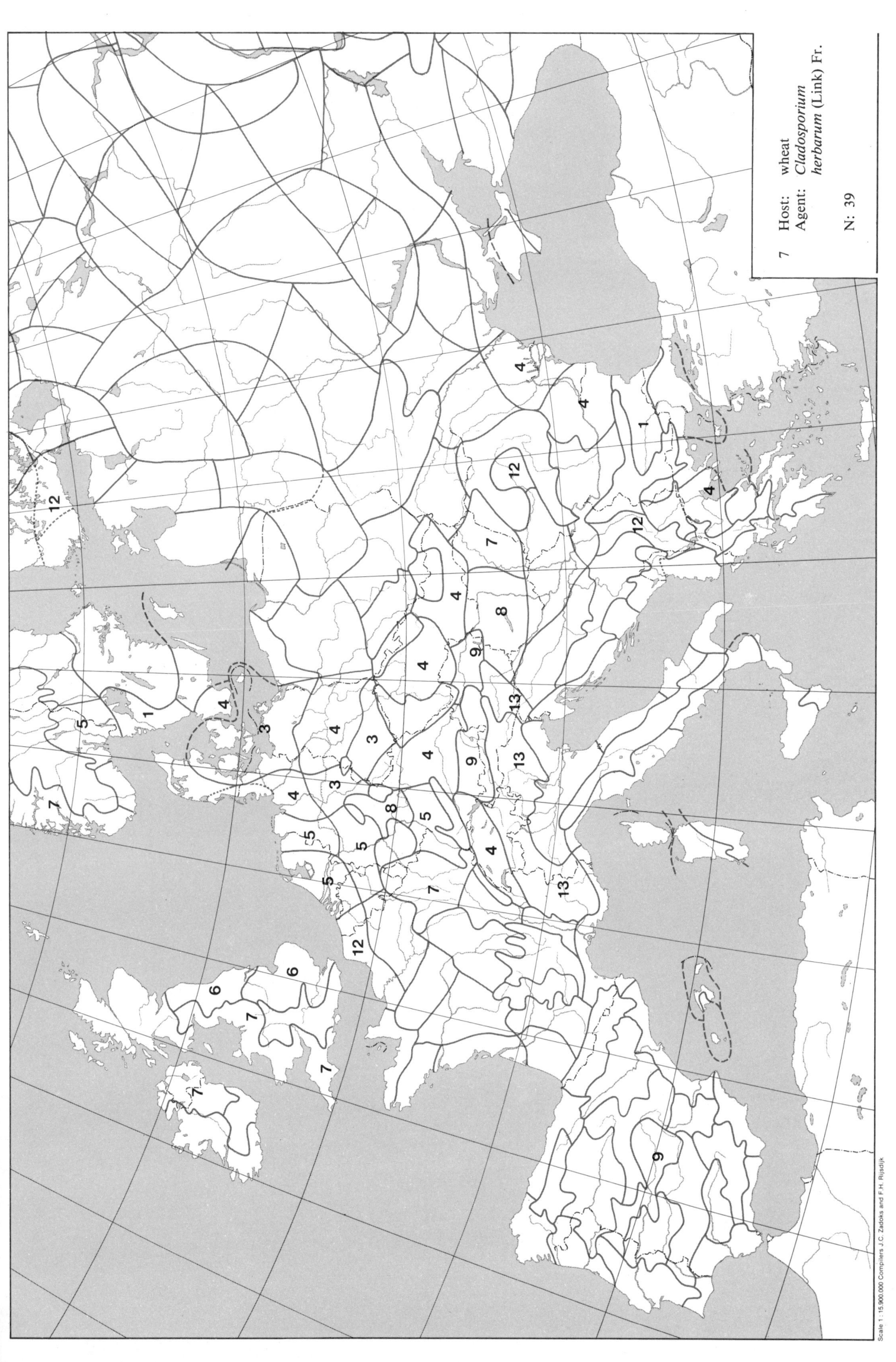

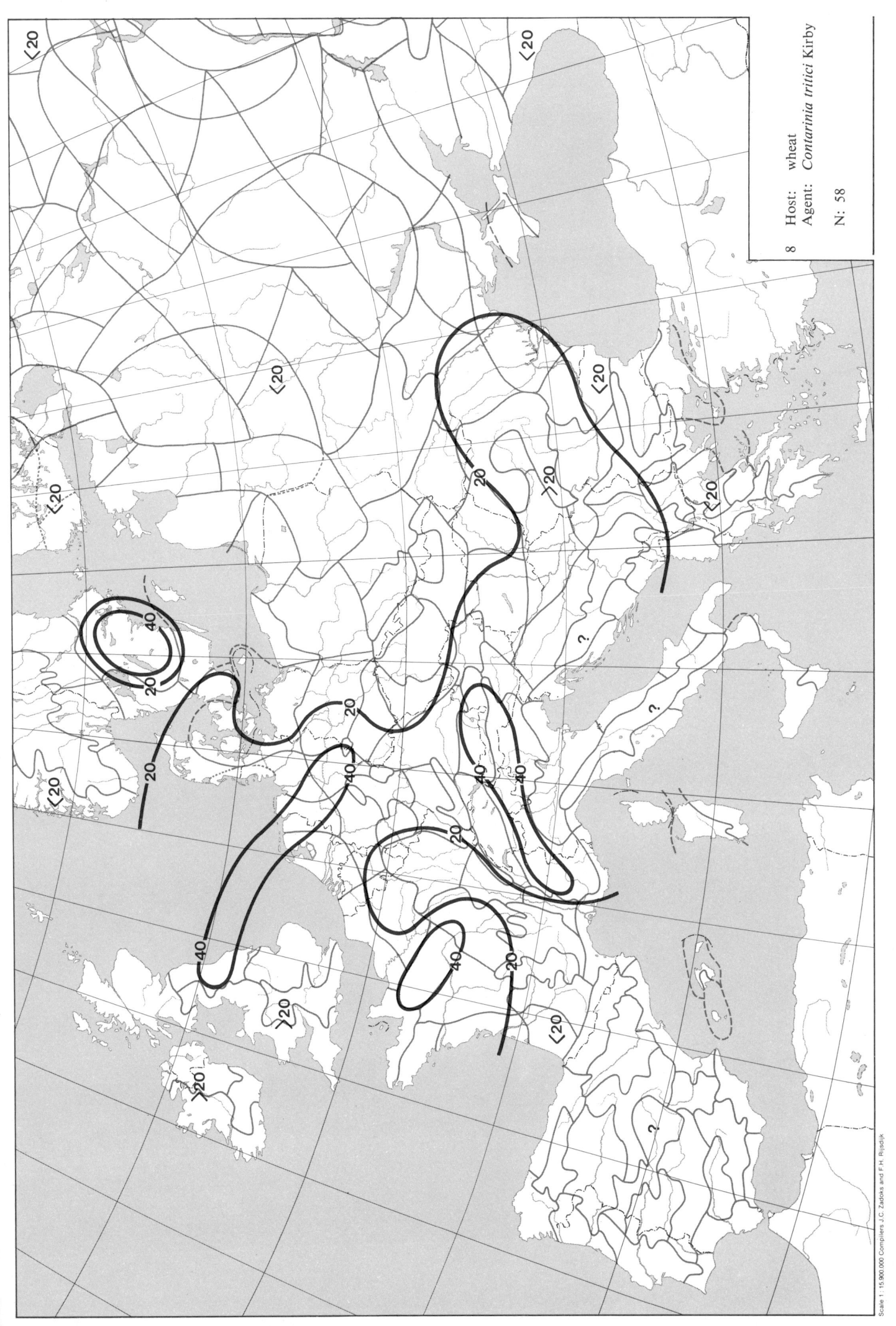

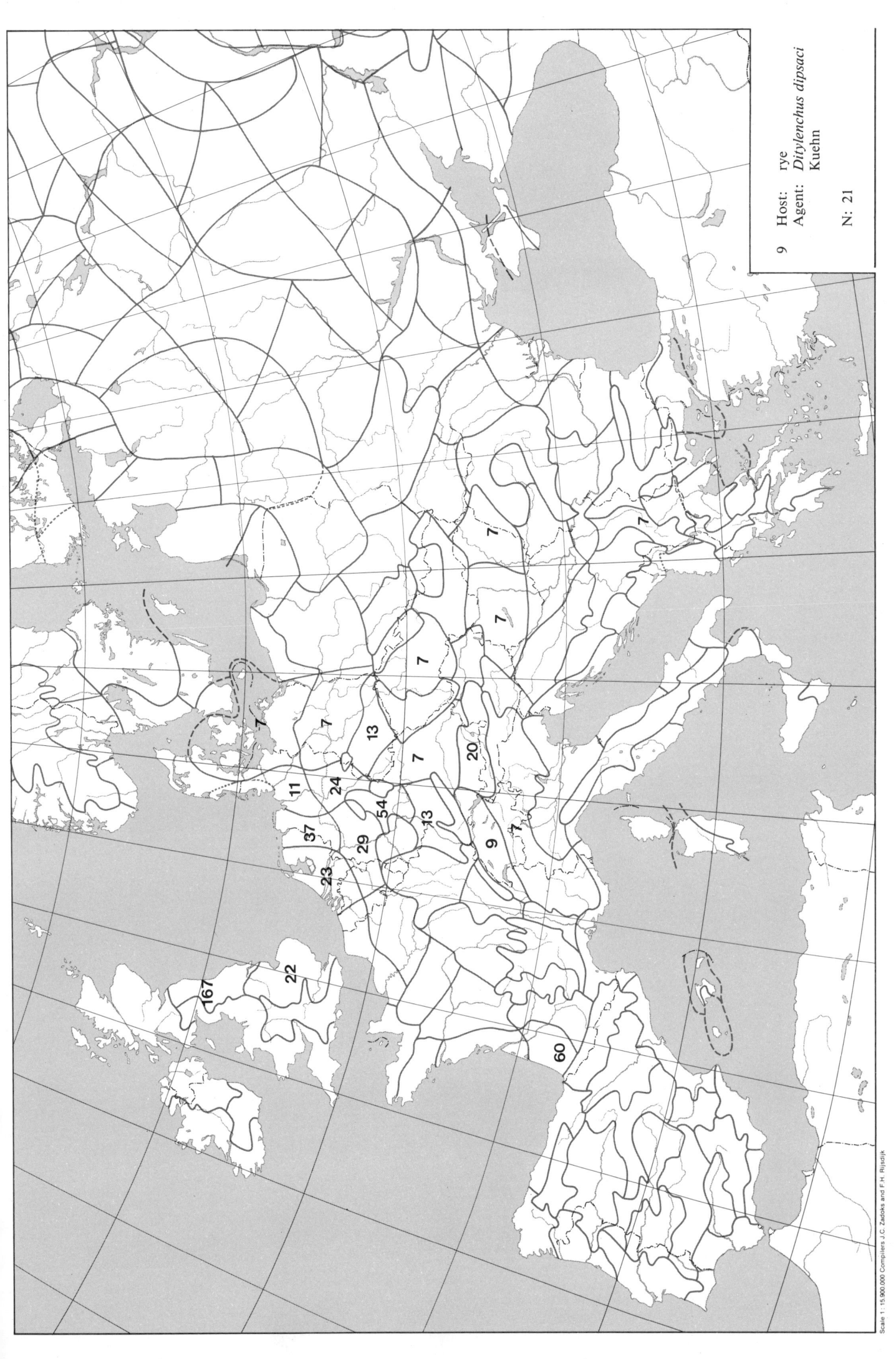

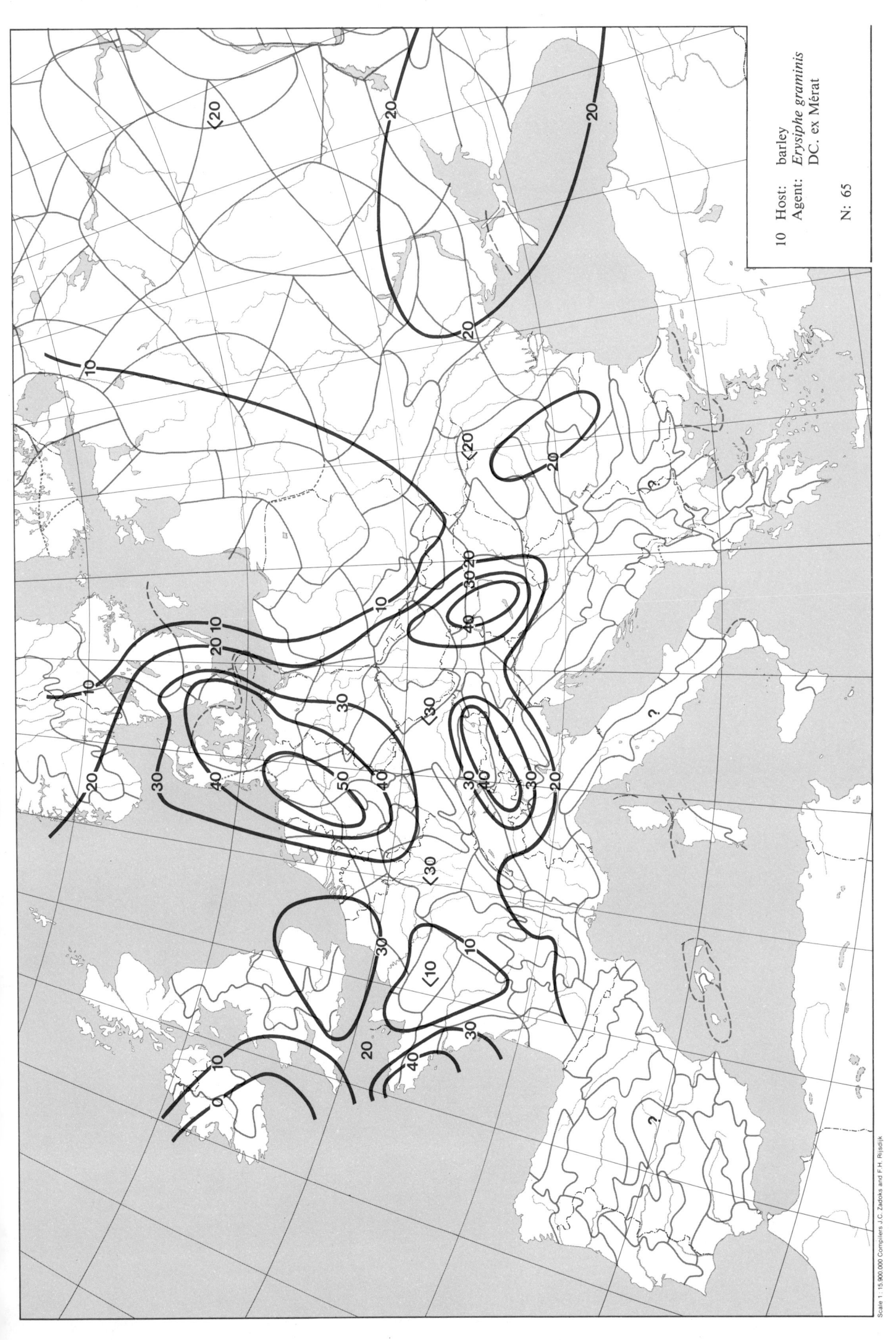

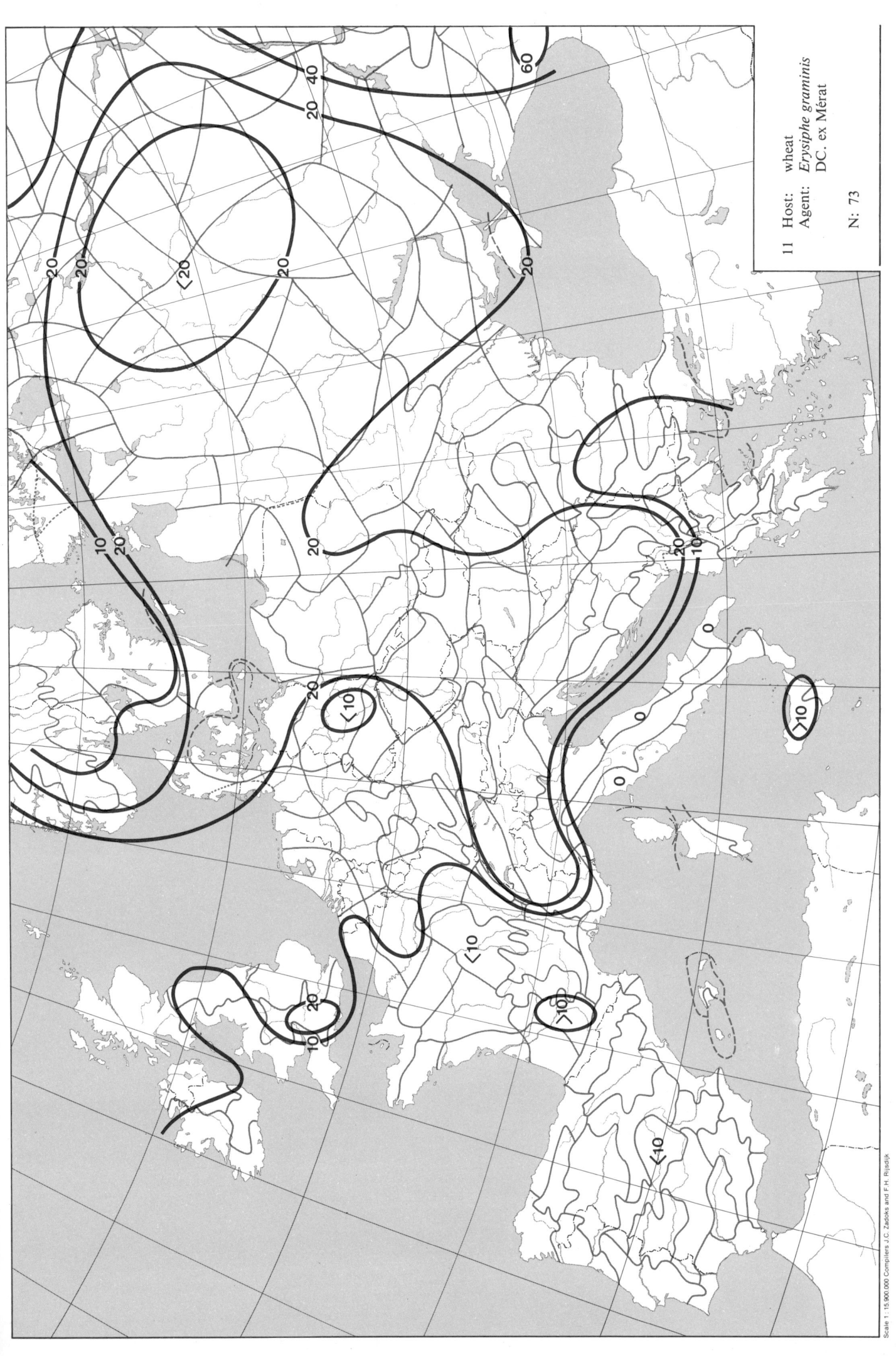

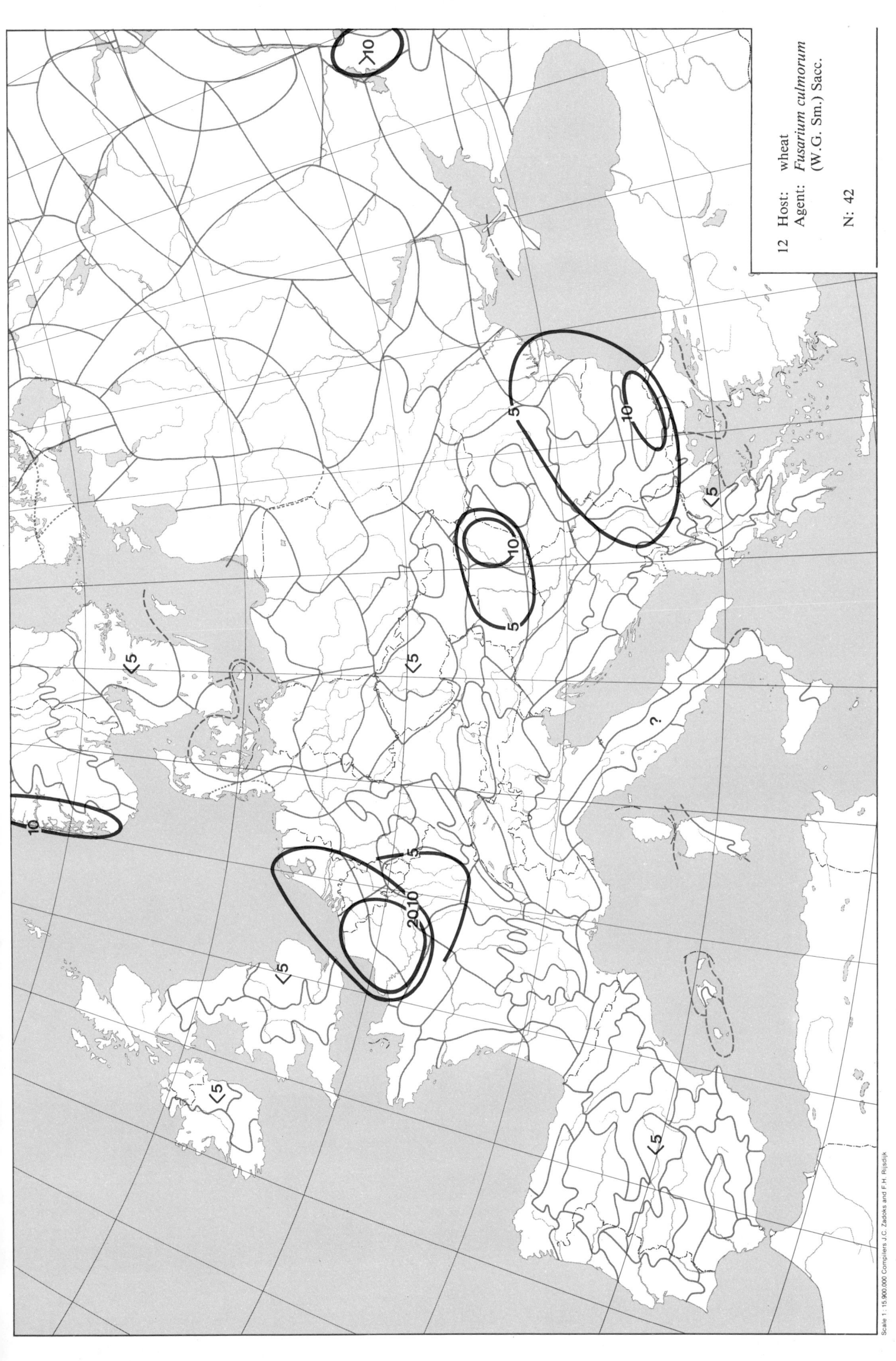

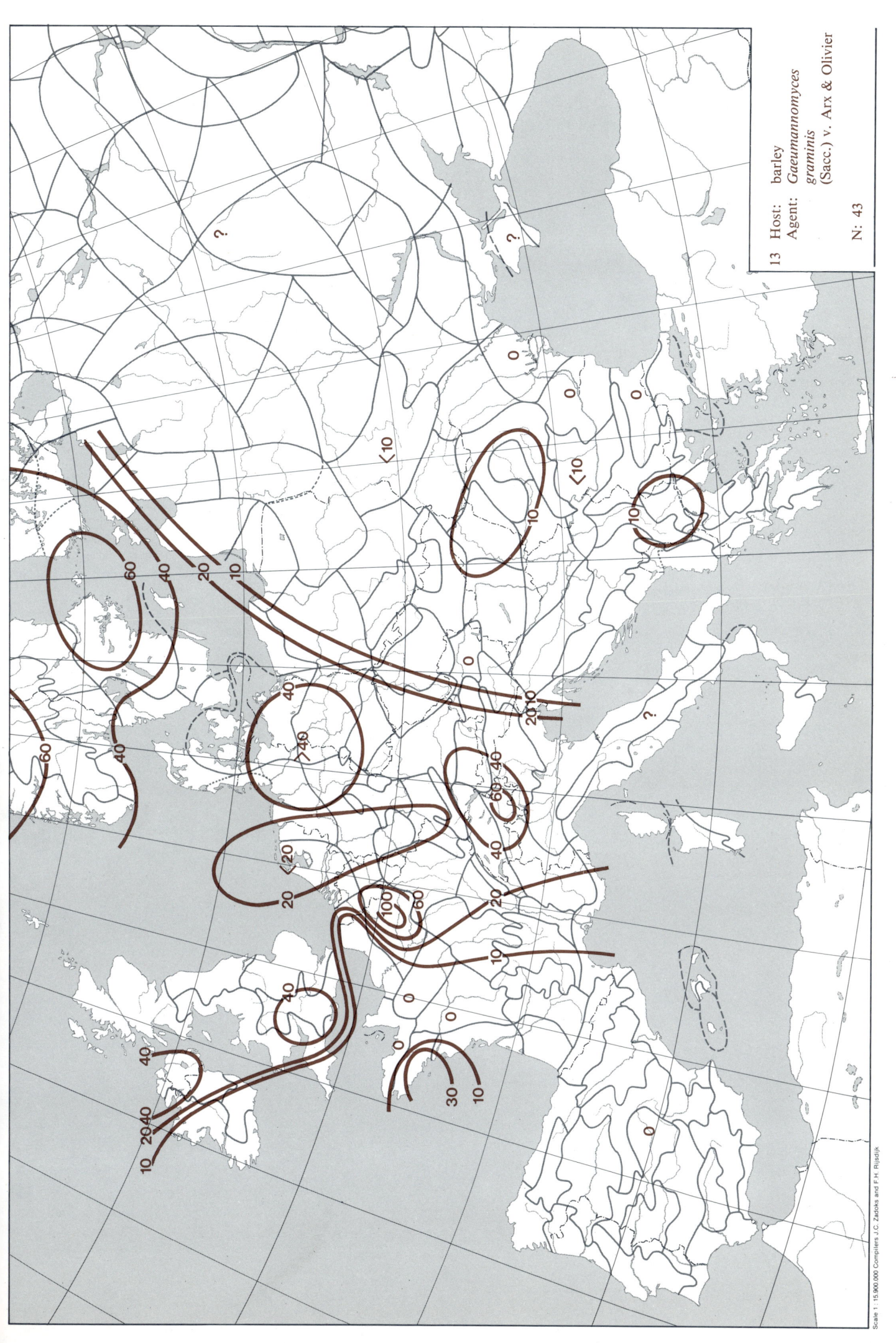

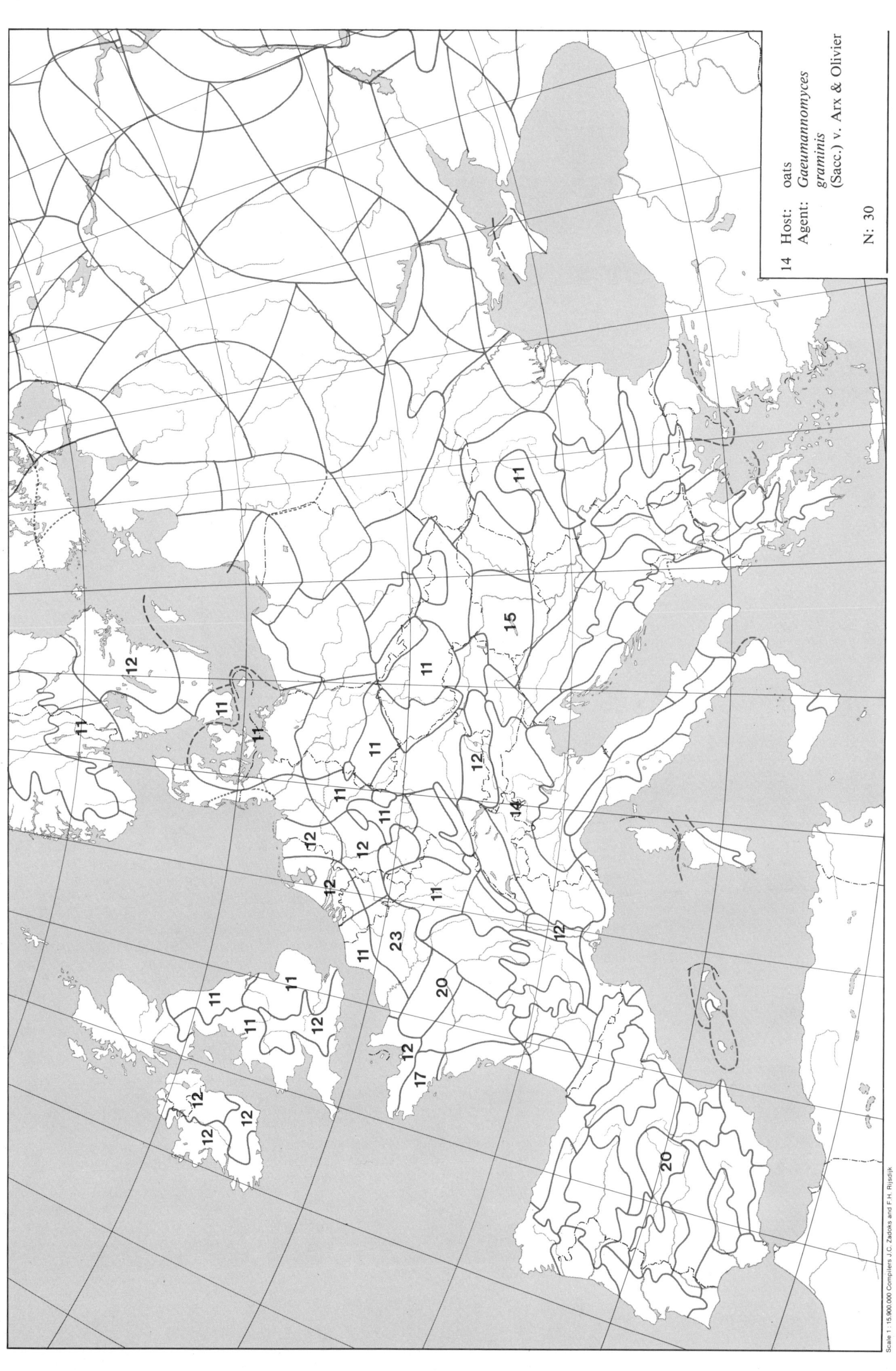

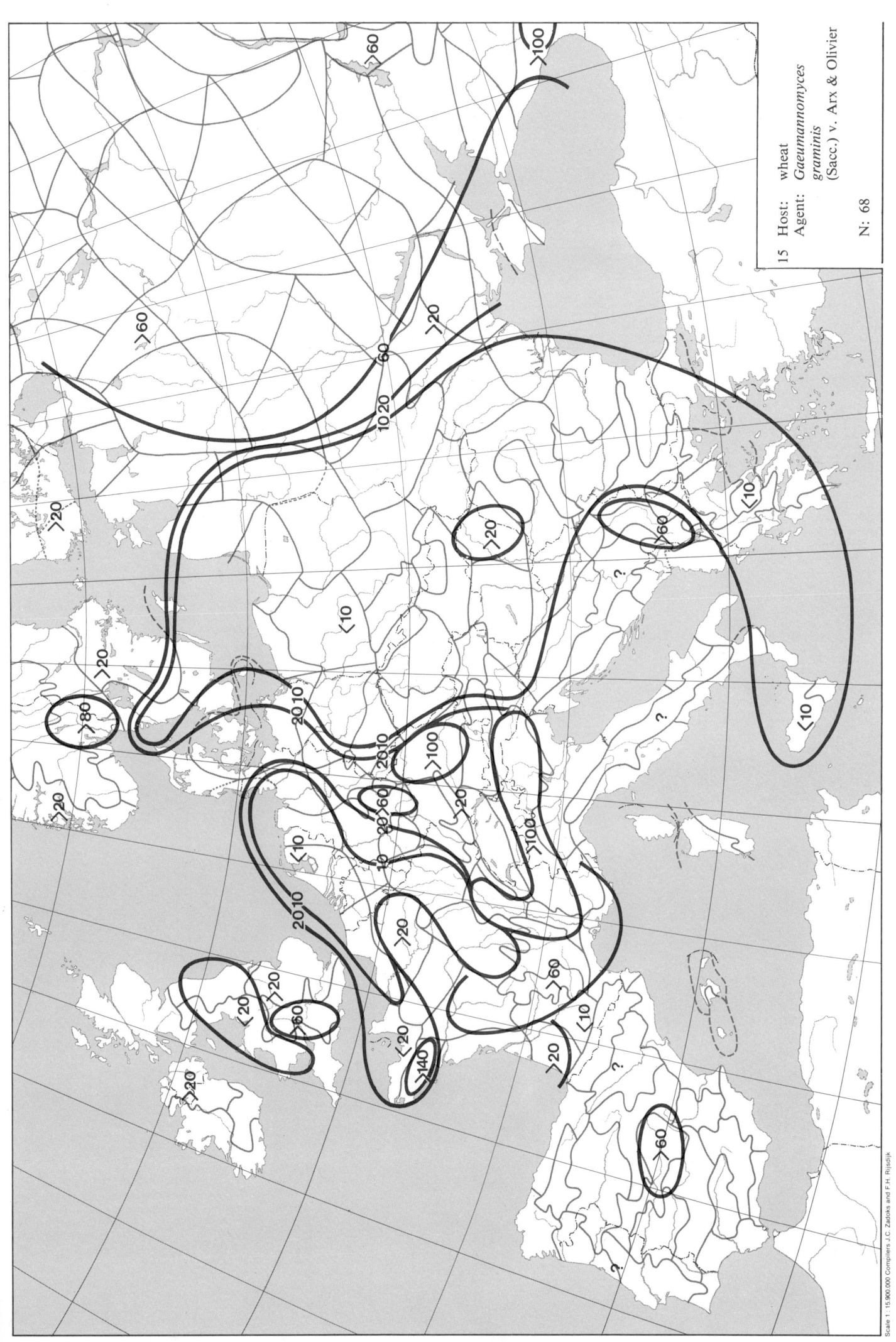

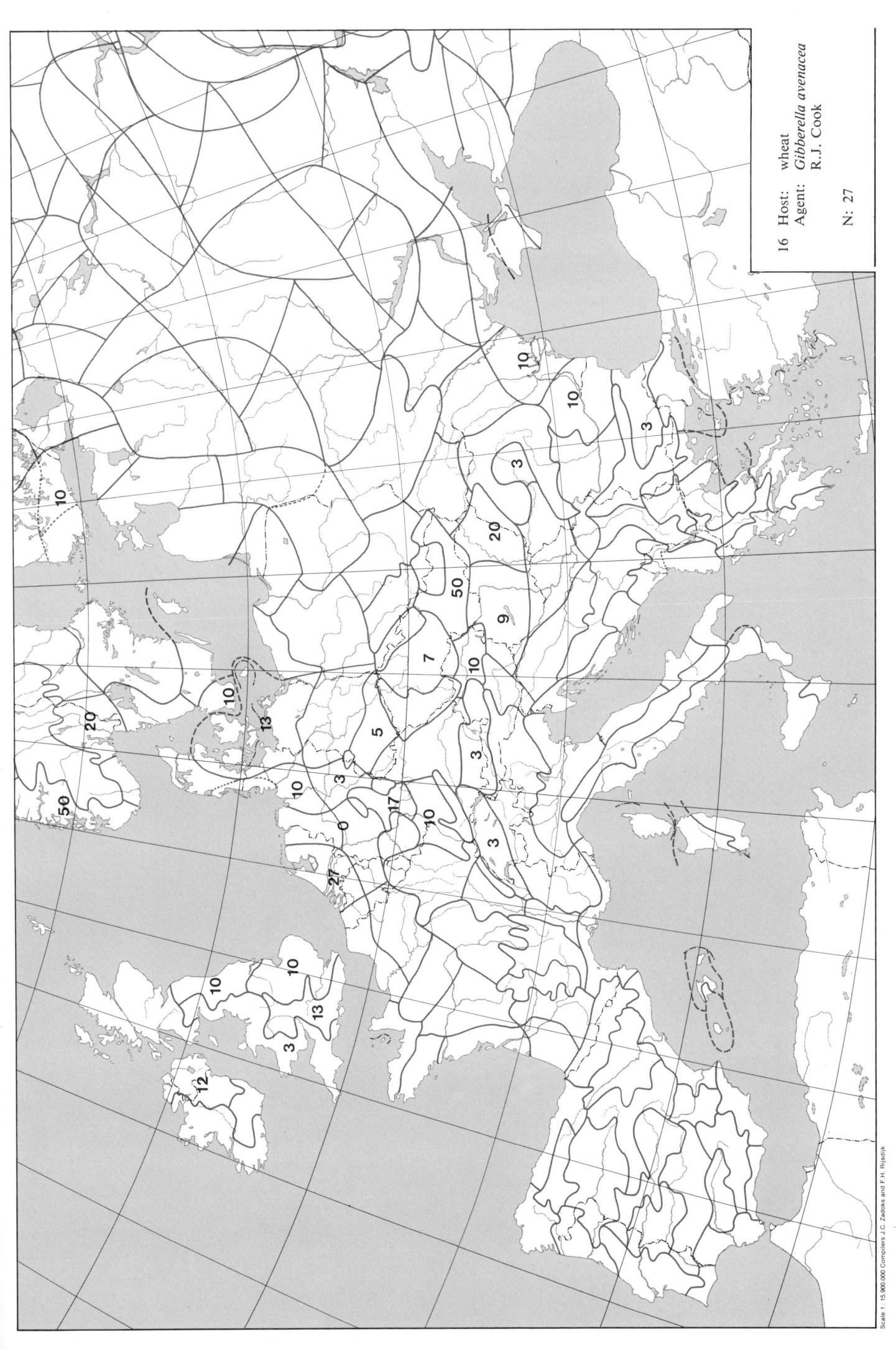

16 Host: wheat
Agent: *Gibberella avenacea*
R.J. Cook

N: 27

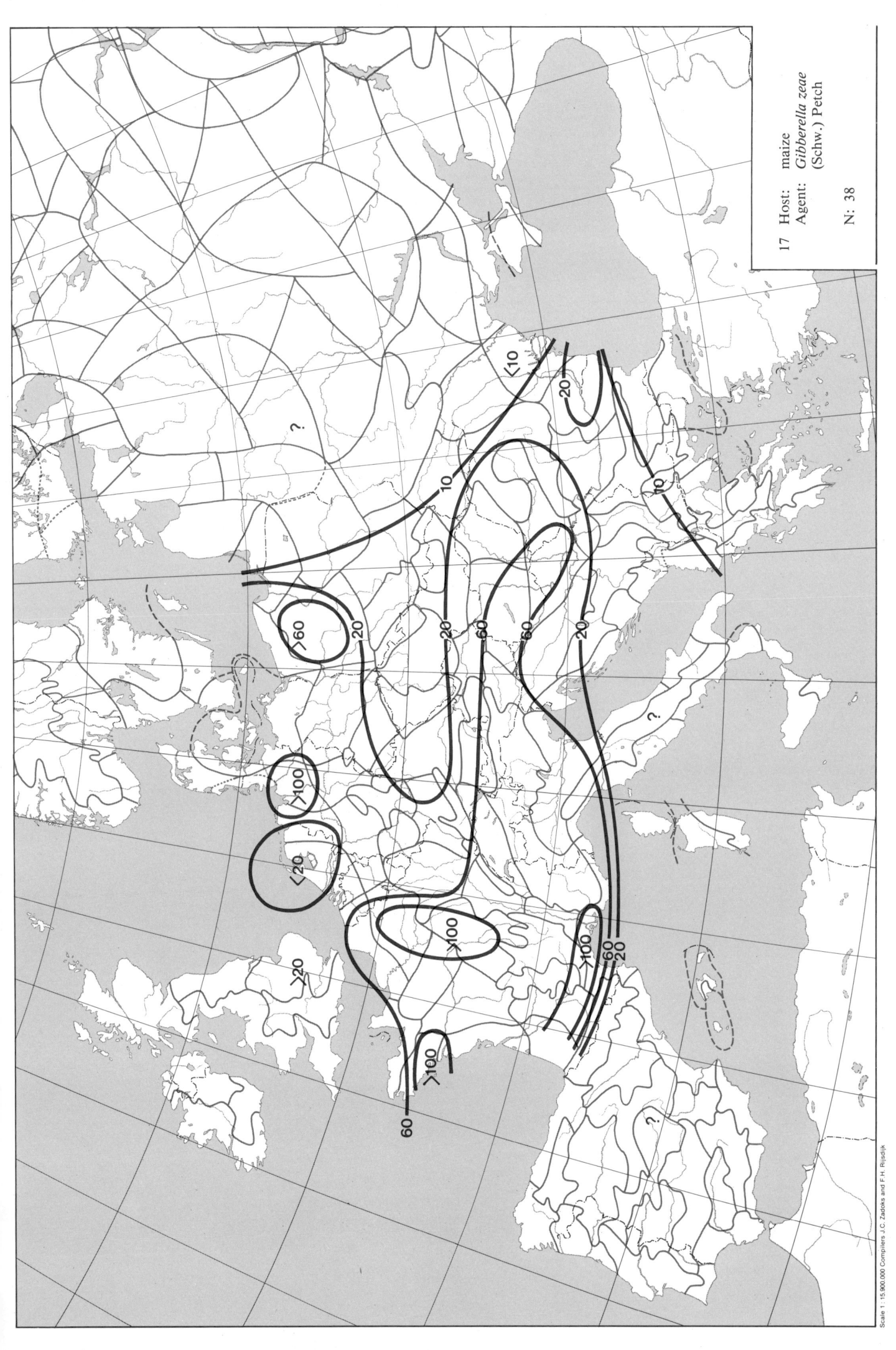

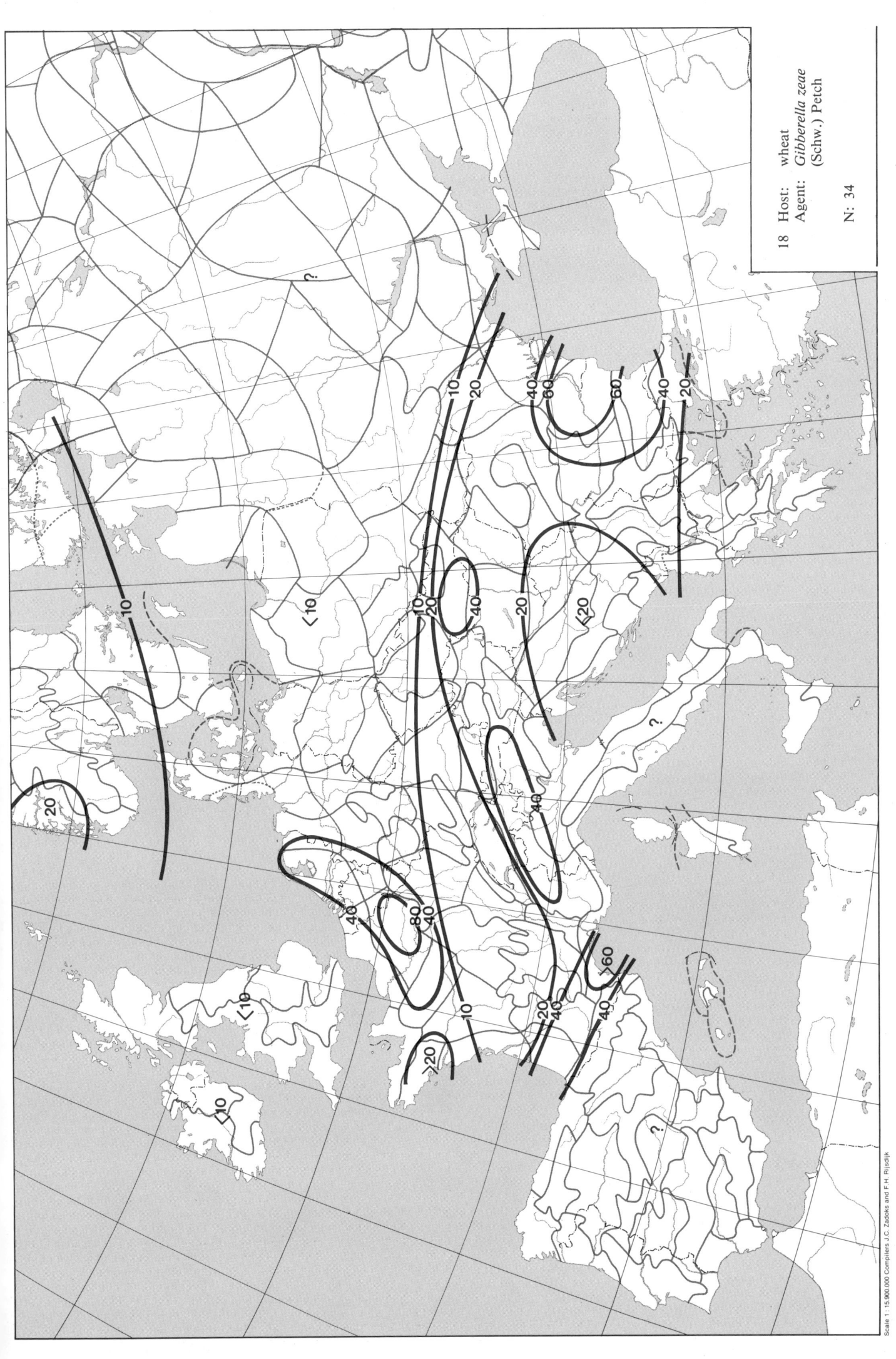

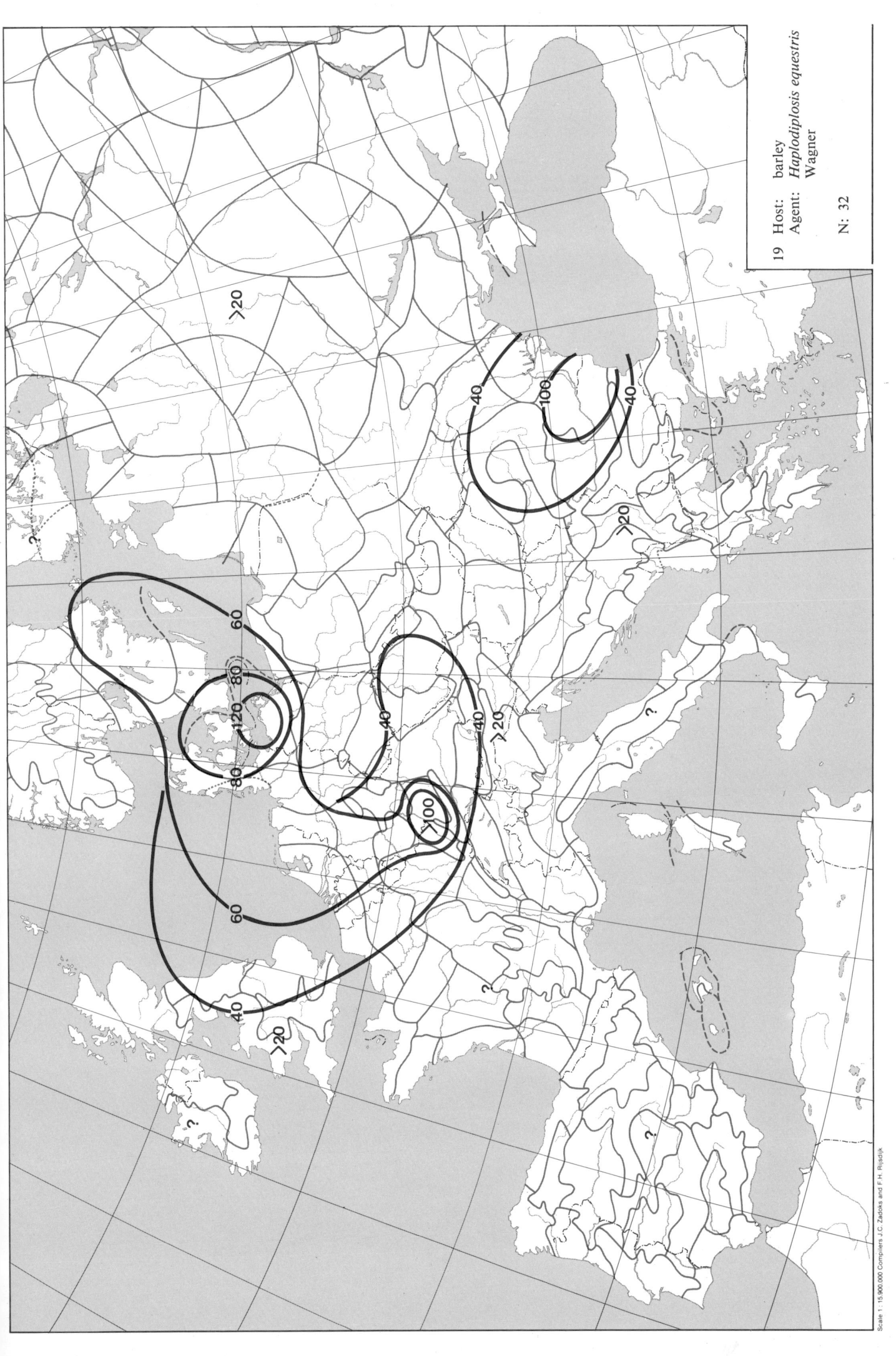

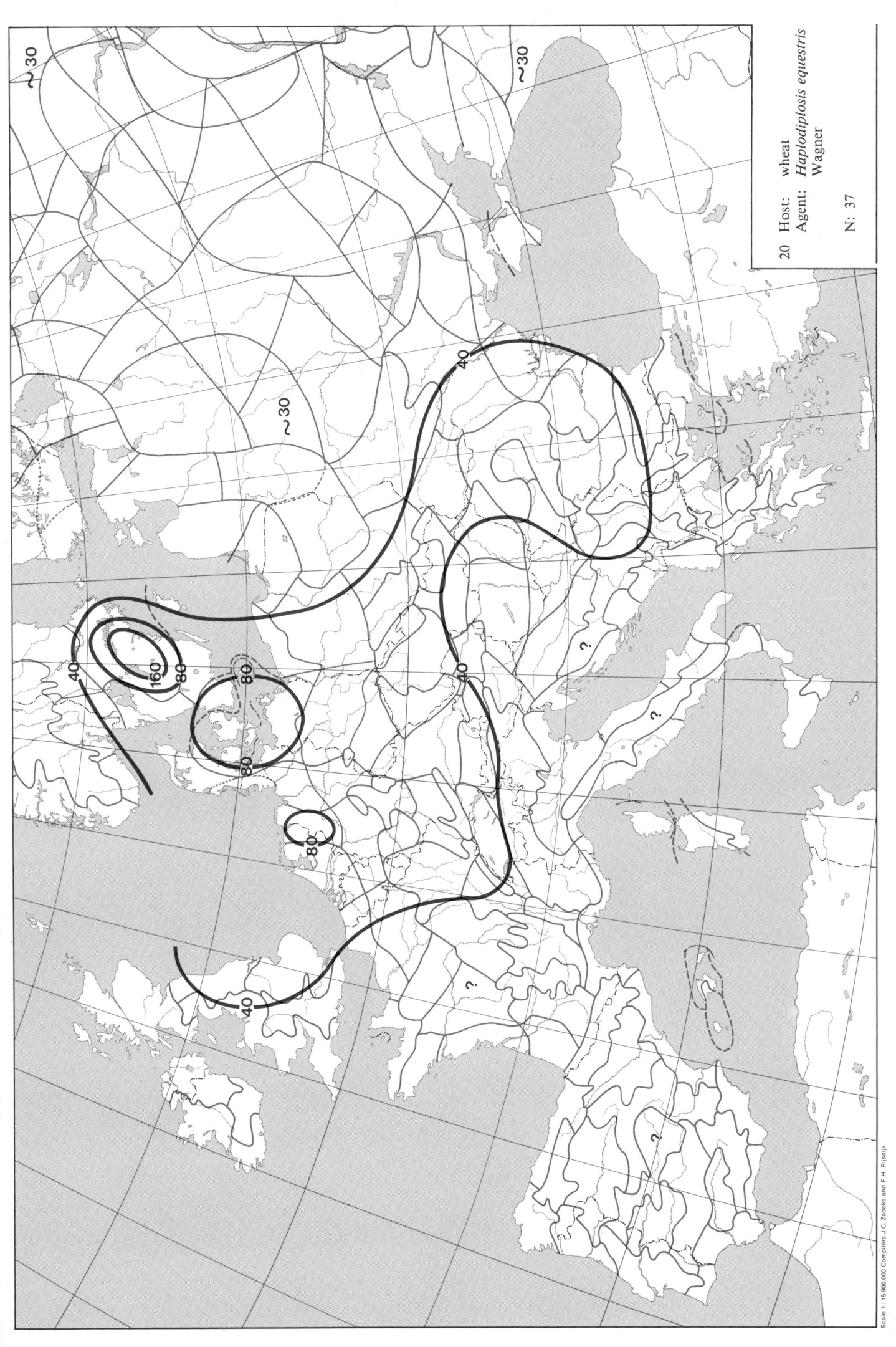

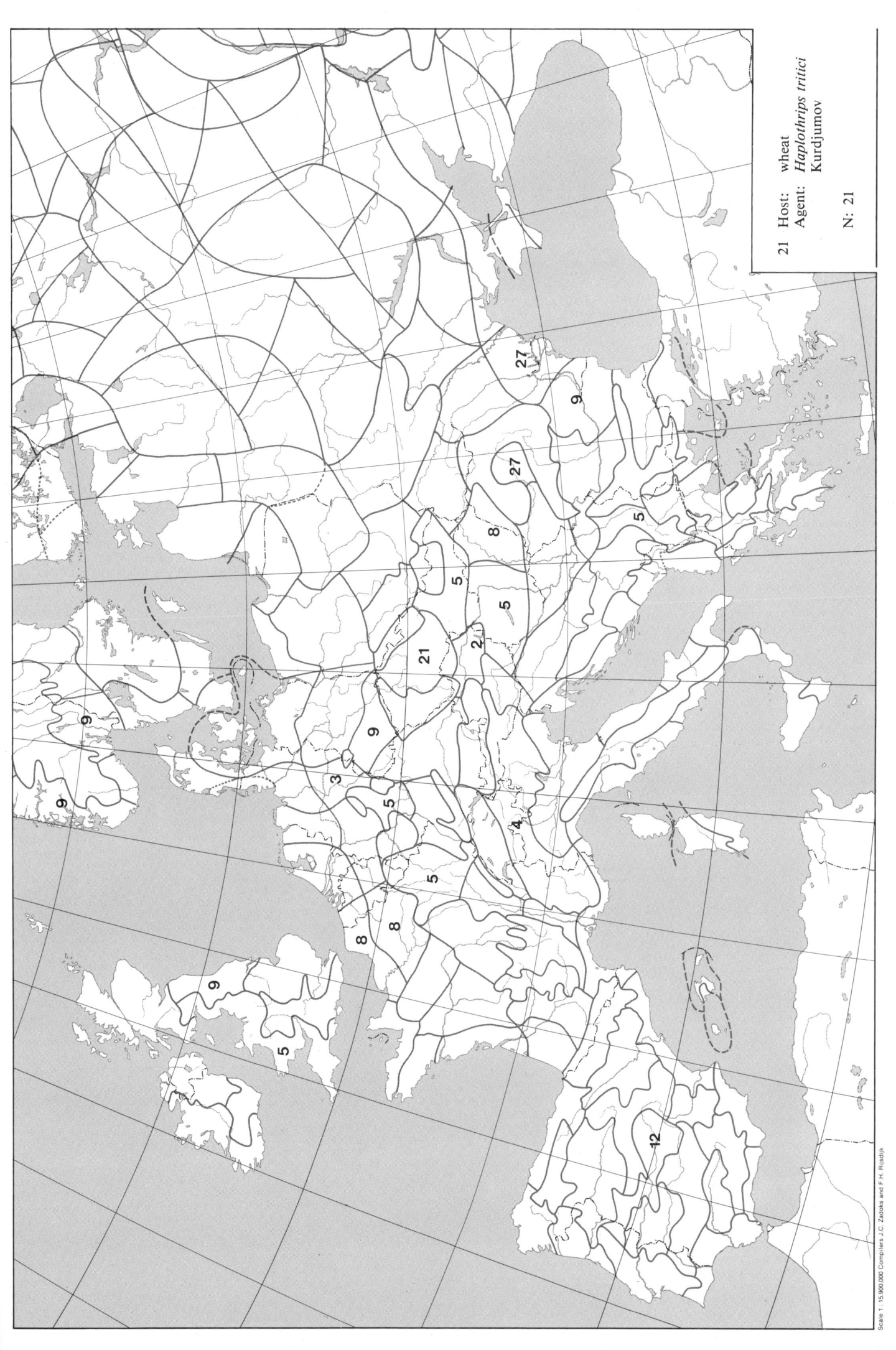

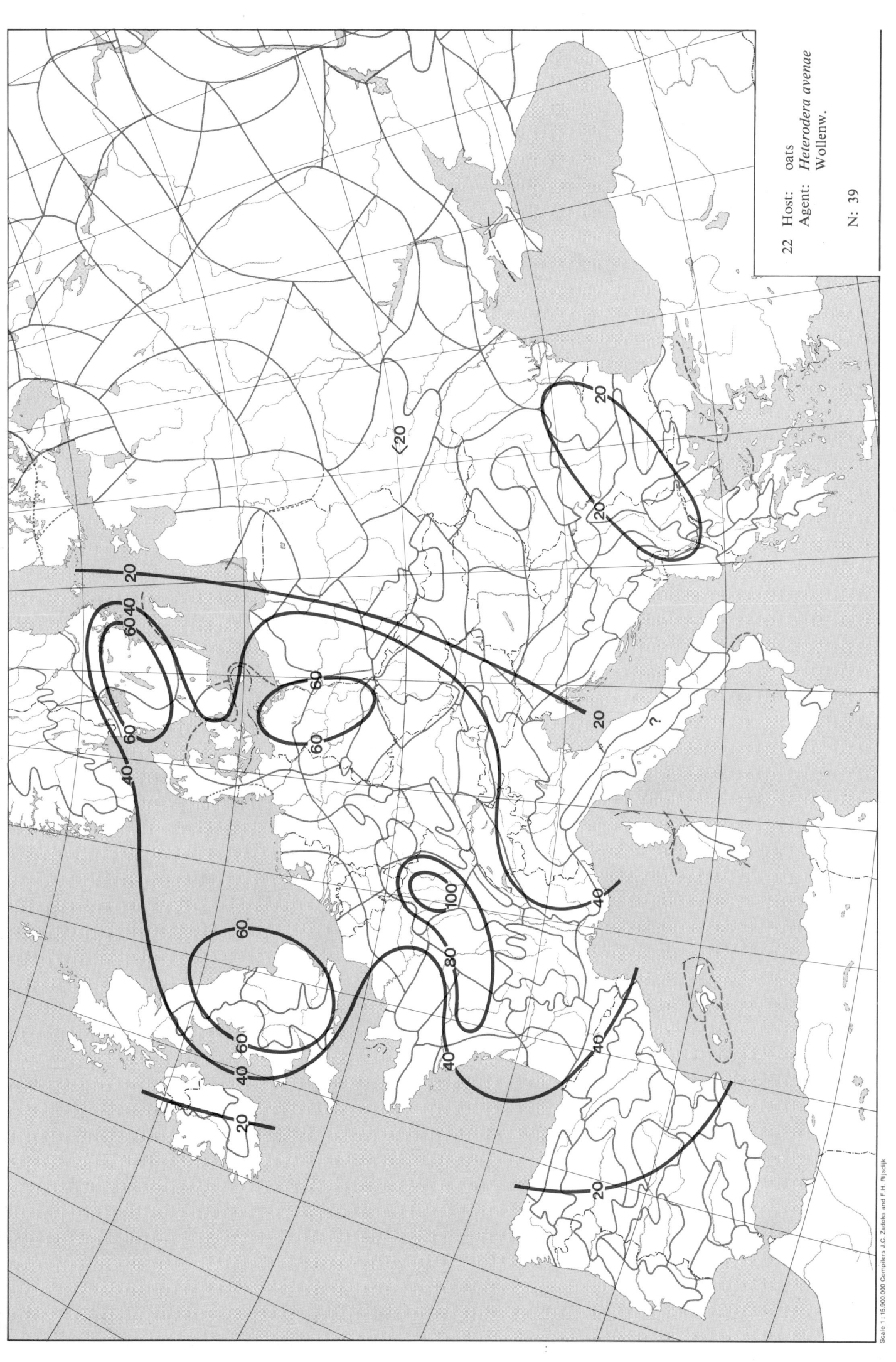

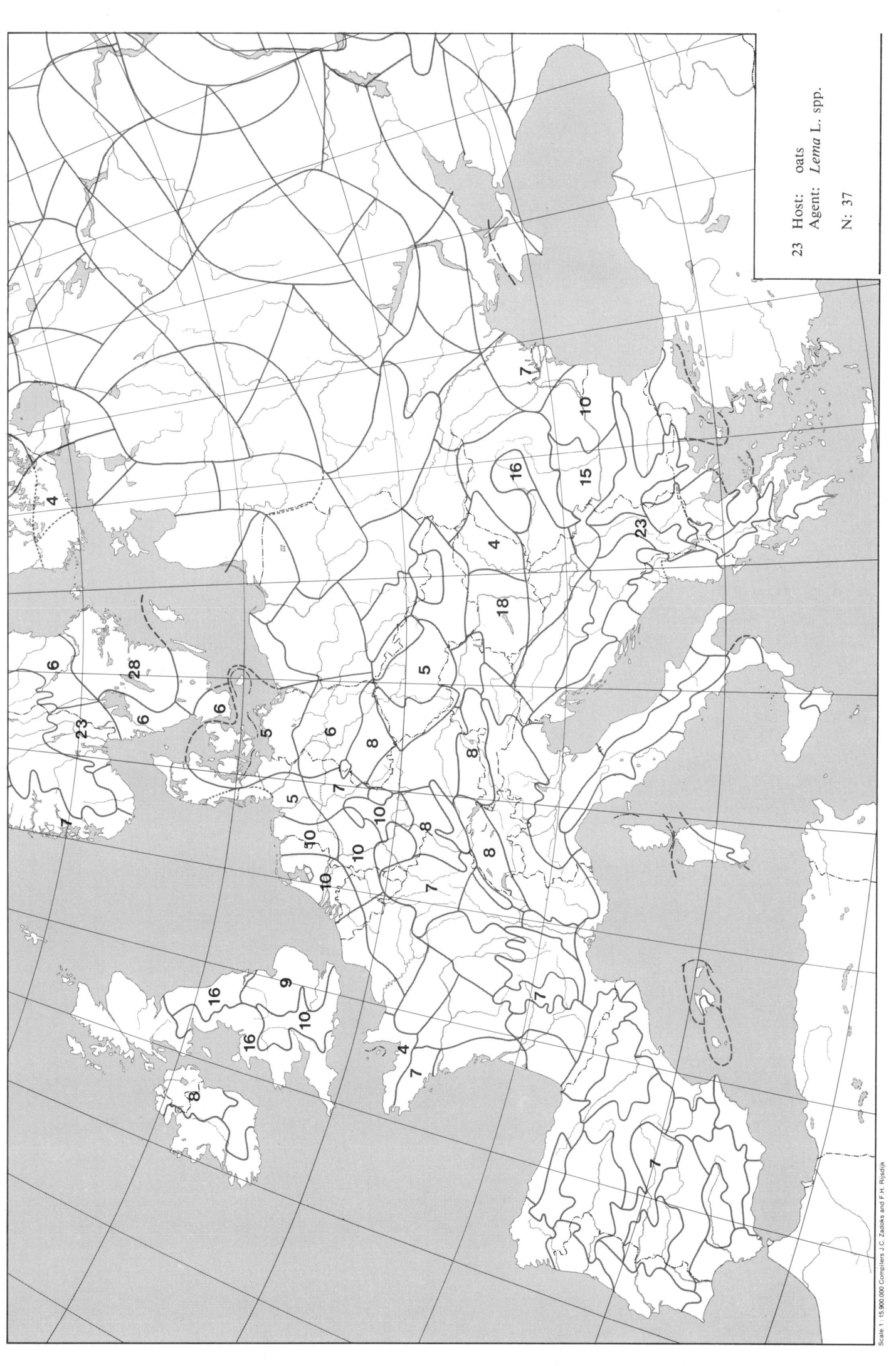

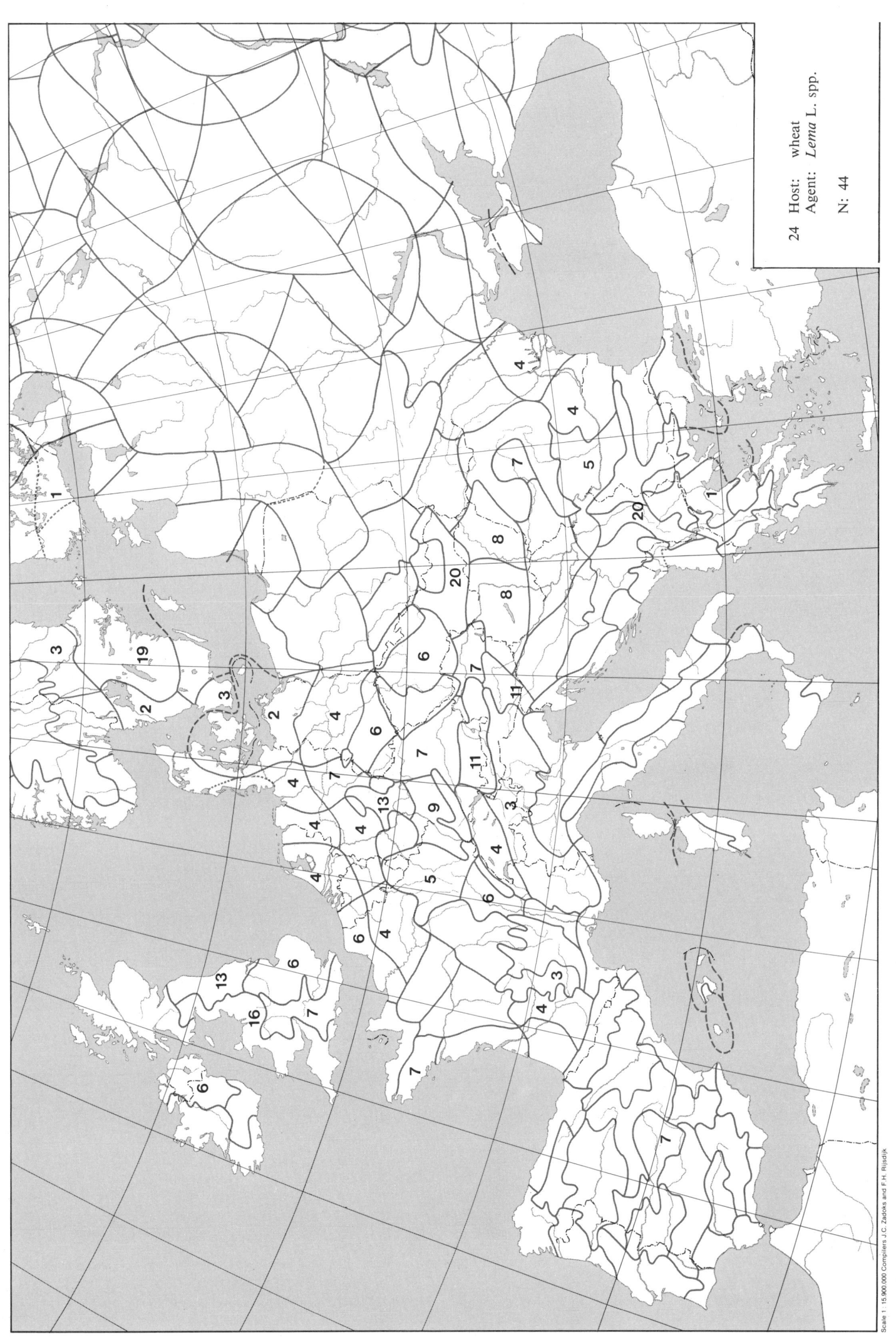

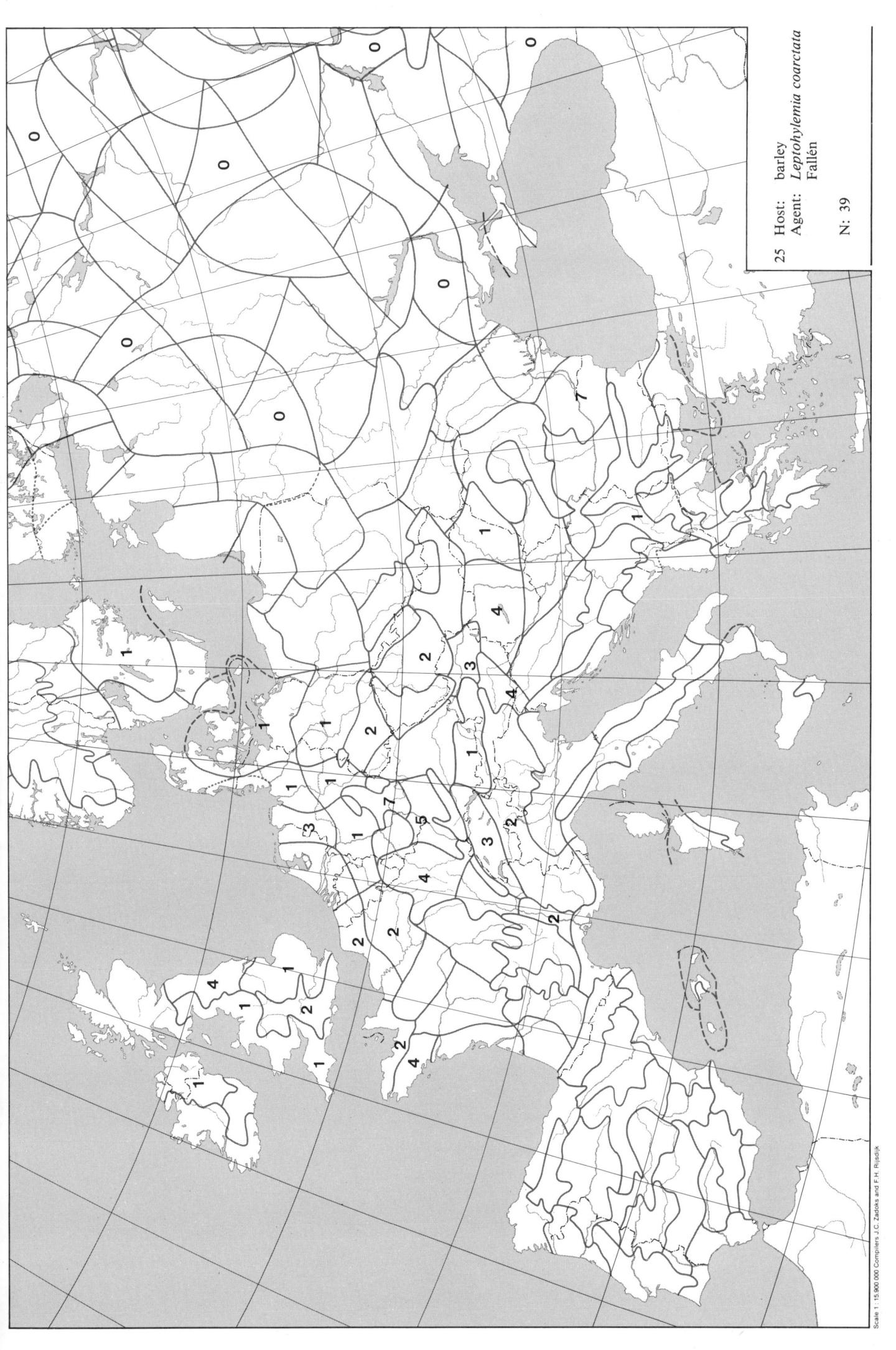

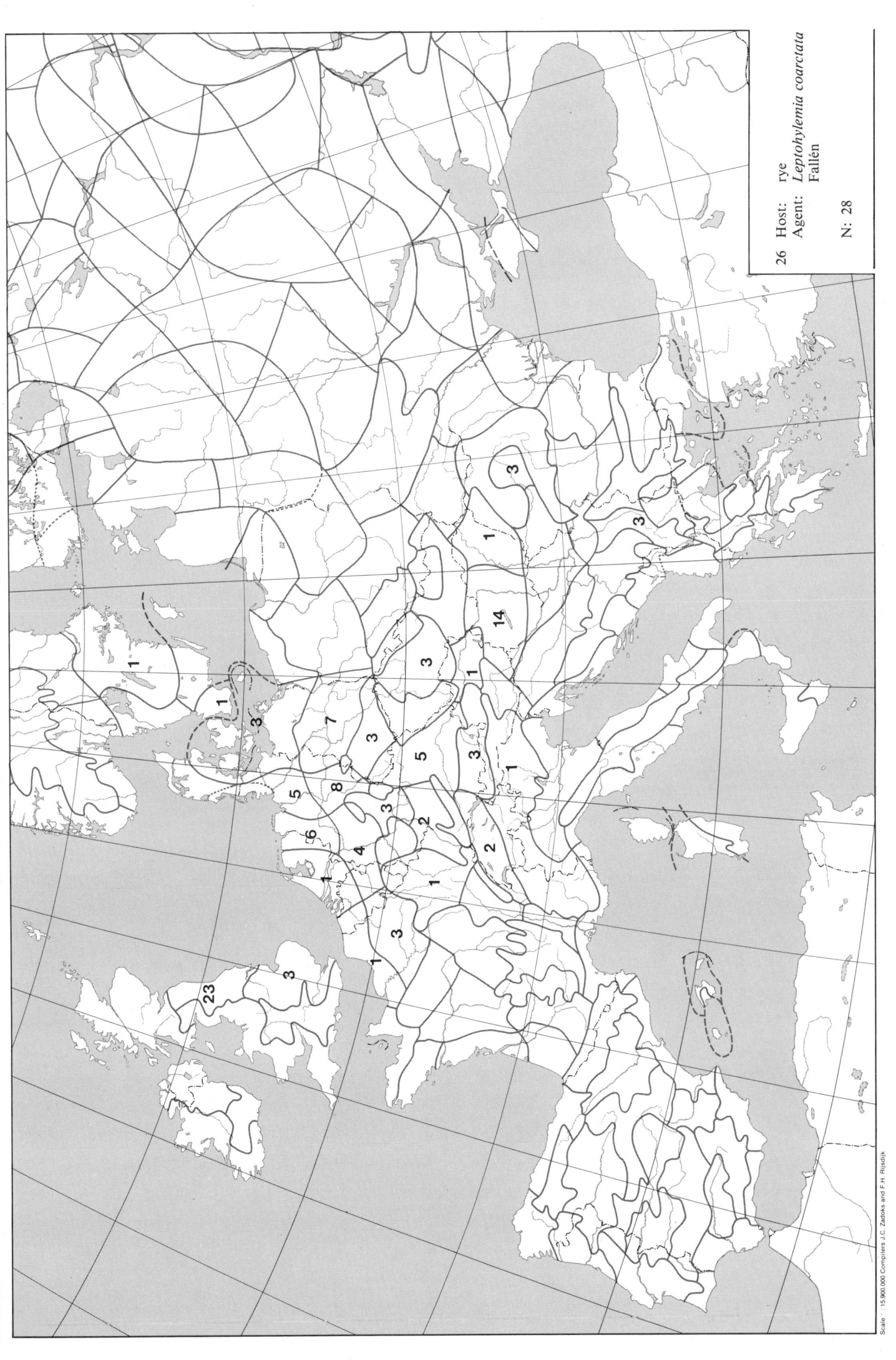

26 Host: rye
Agent: *Leptohylemia coarctata* Fallén

N: 28

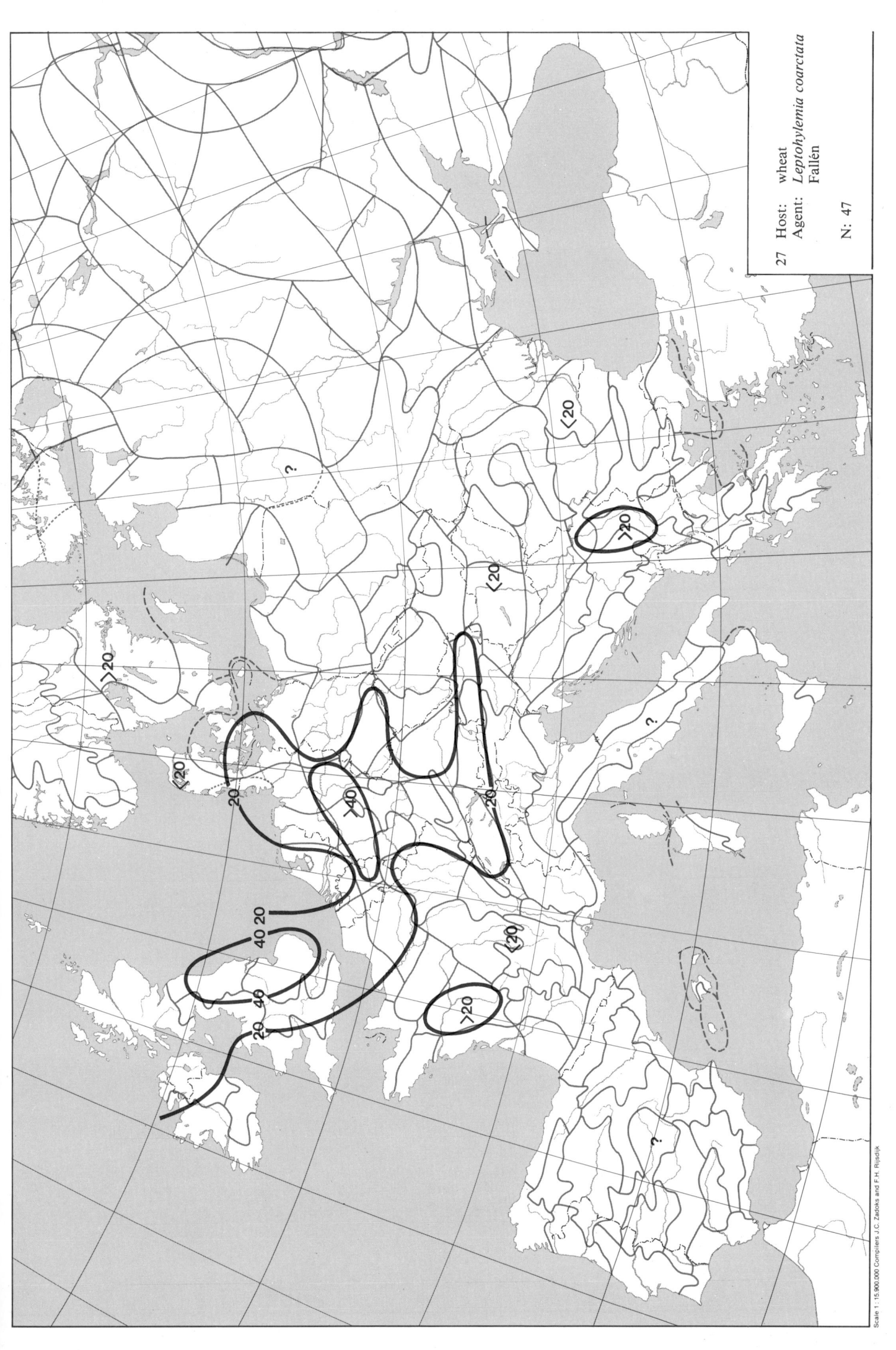

27 Host: wheat
Agent: *Leptohylemia coarctata* Fallén

N: 47

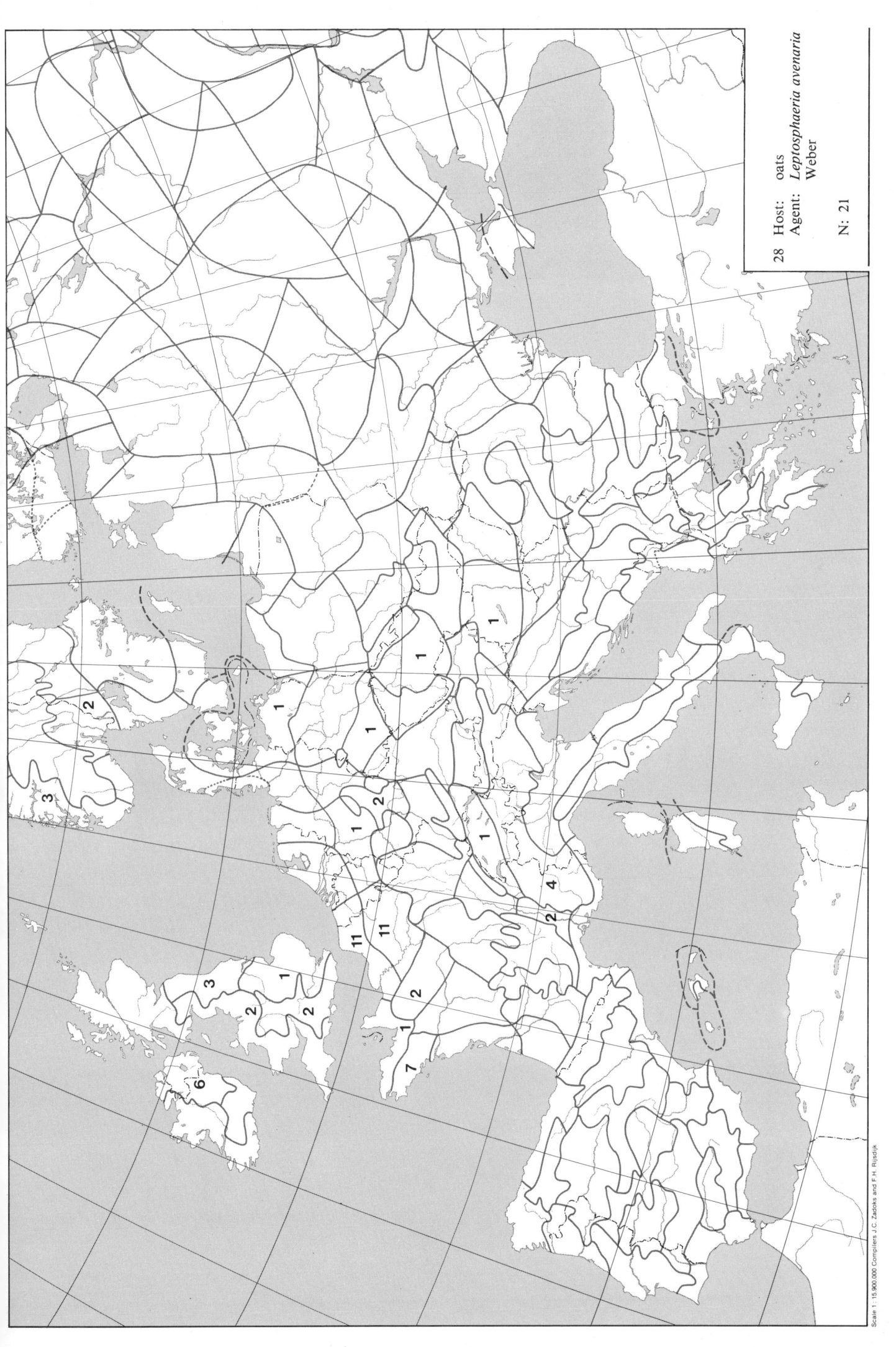

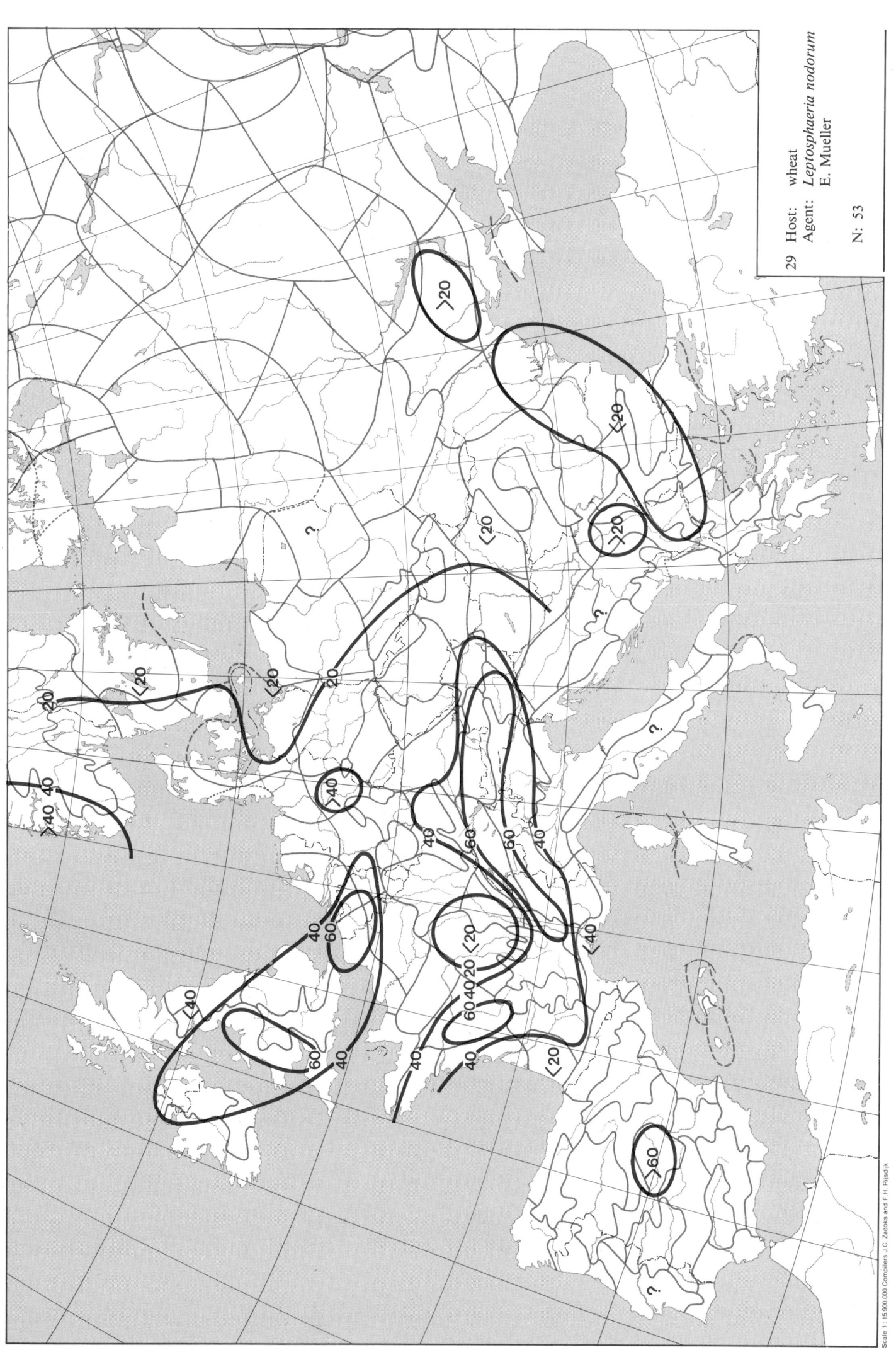

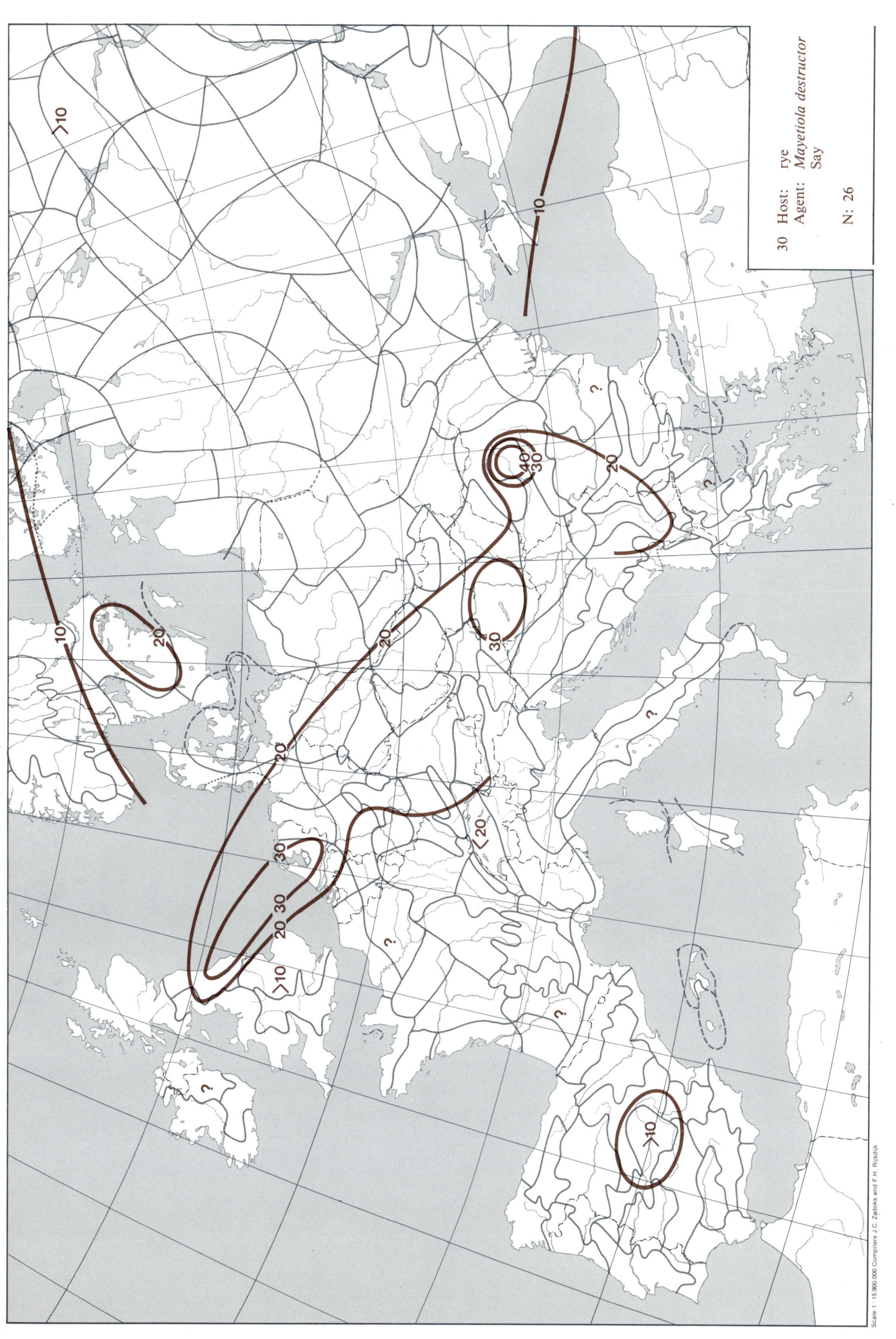

30 Host: rye
Agent: *Mayetiola destructor* Say
N: 26

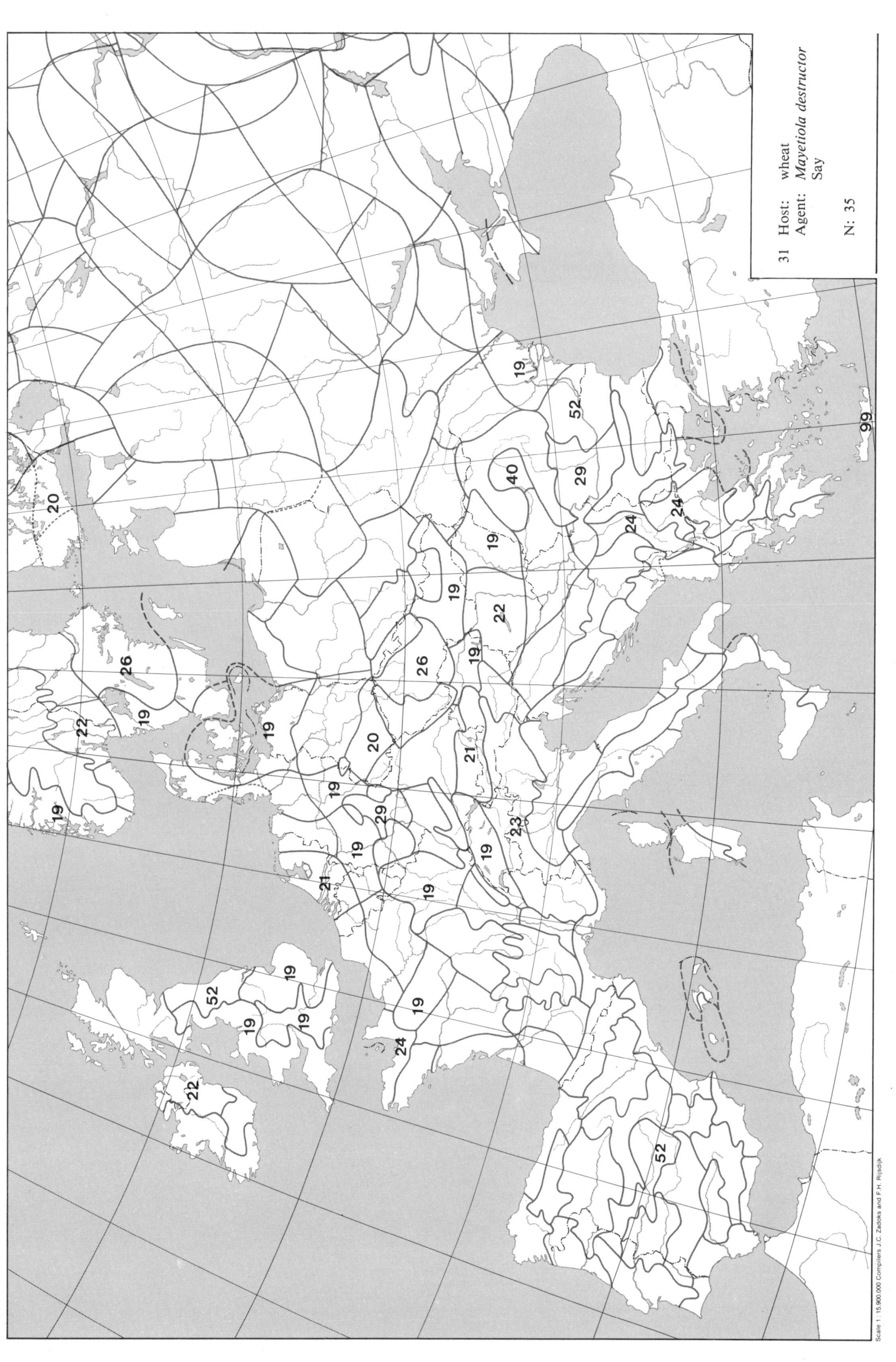

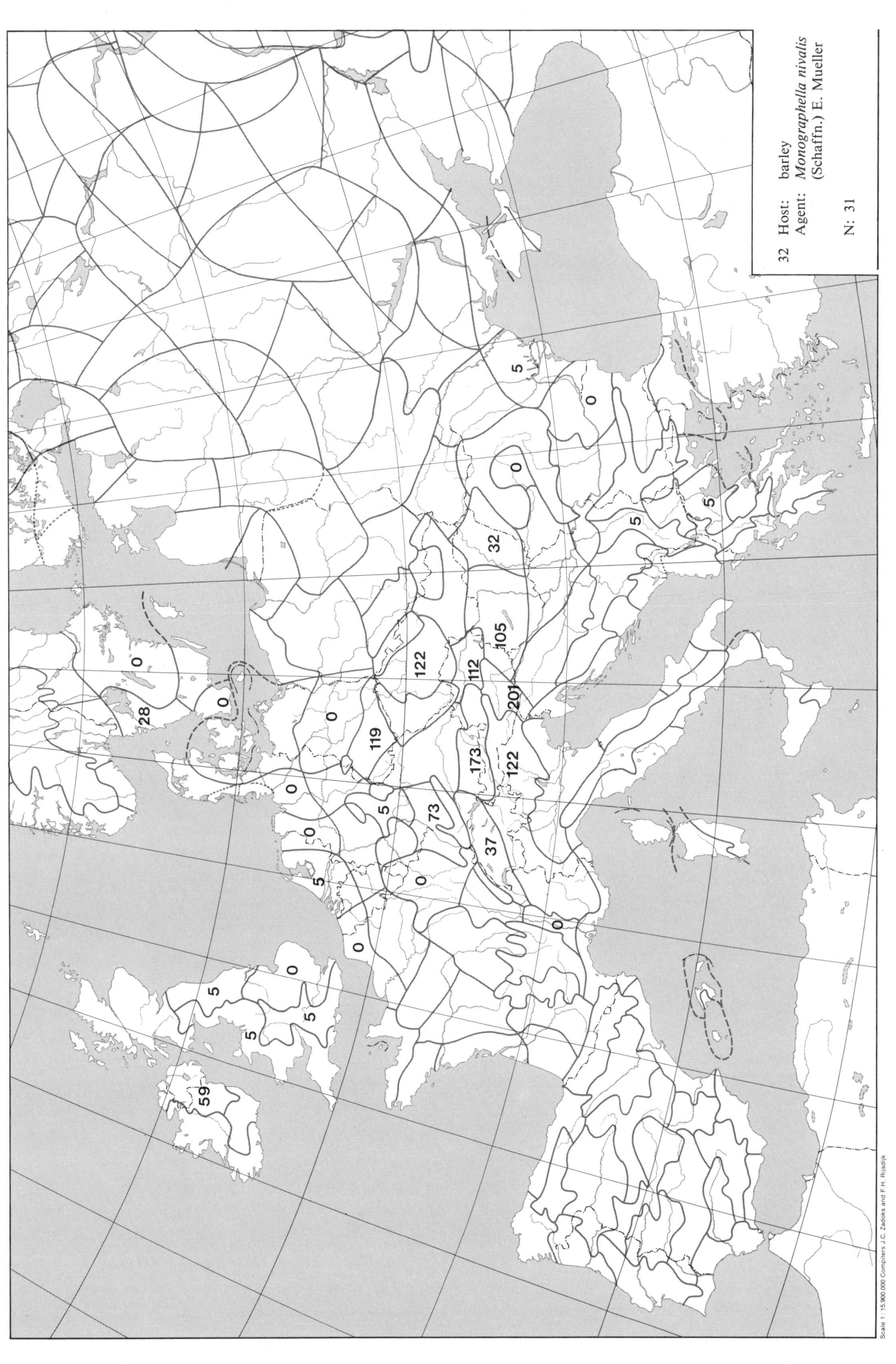

32 Host: barley
Agent: *Monographella nivalis* (Schaffn.) E. Mueller

N: 31

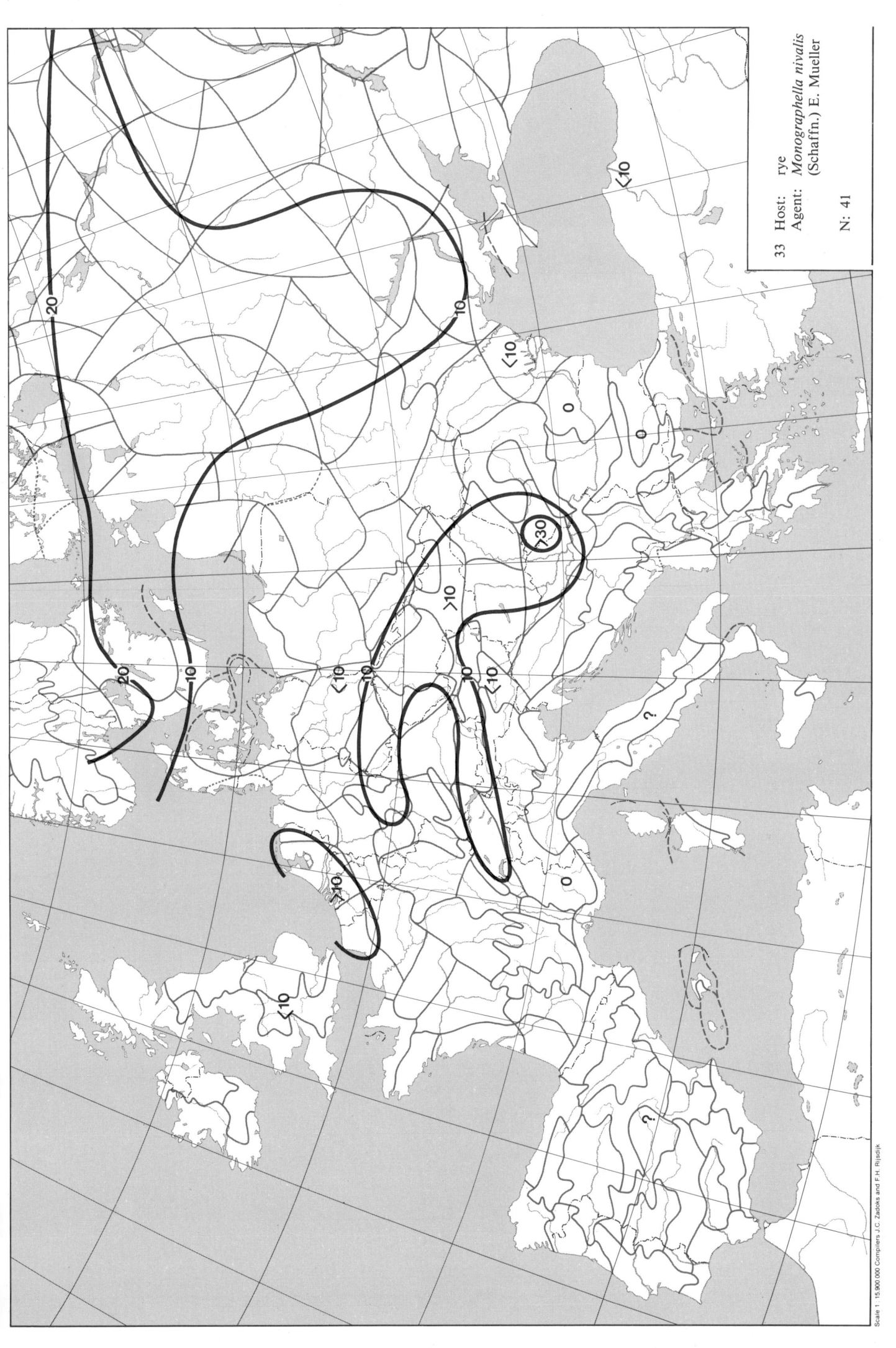

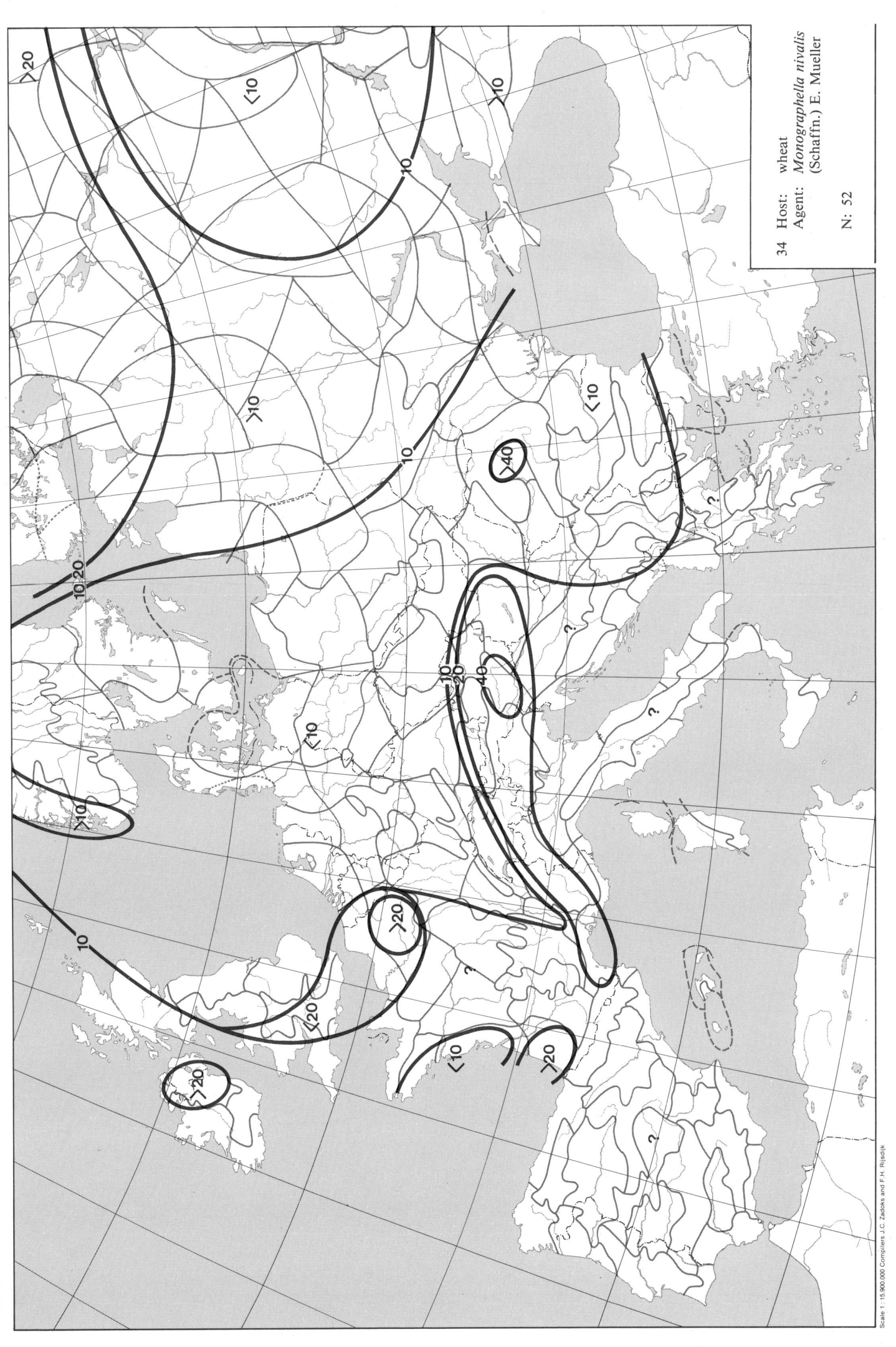

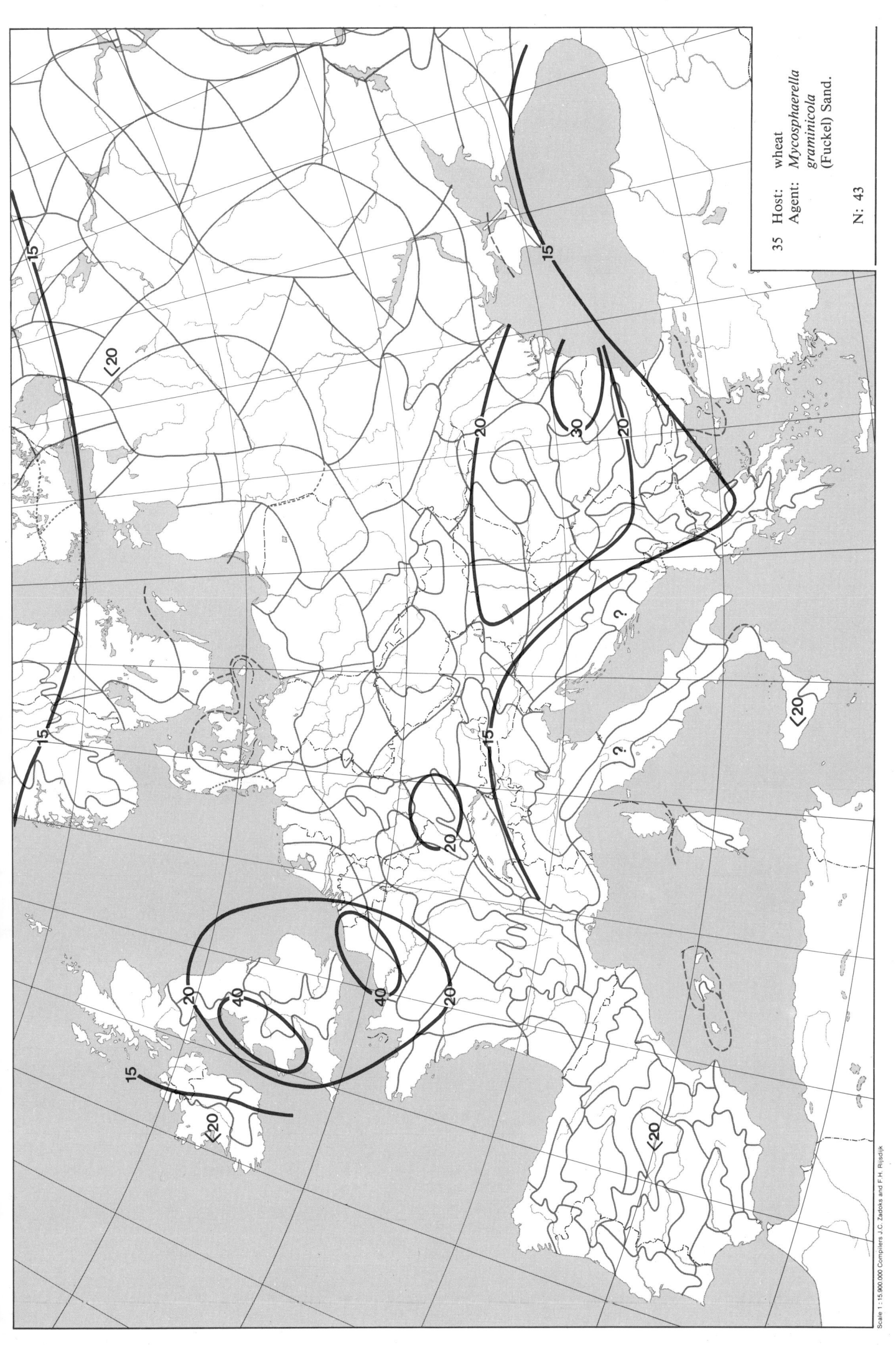

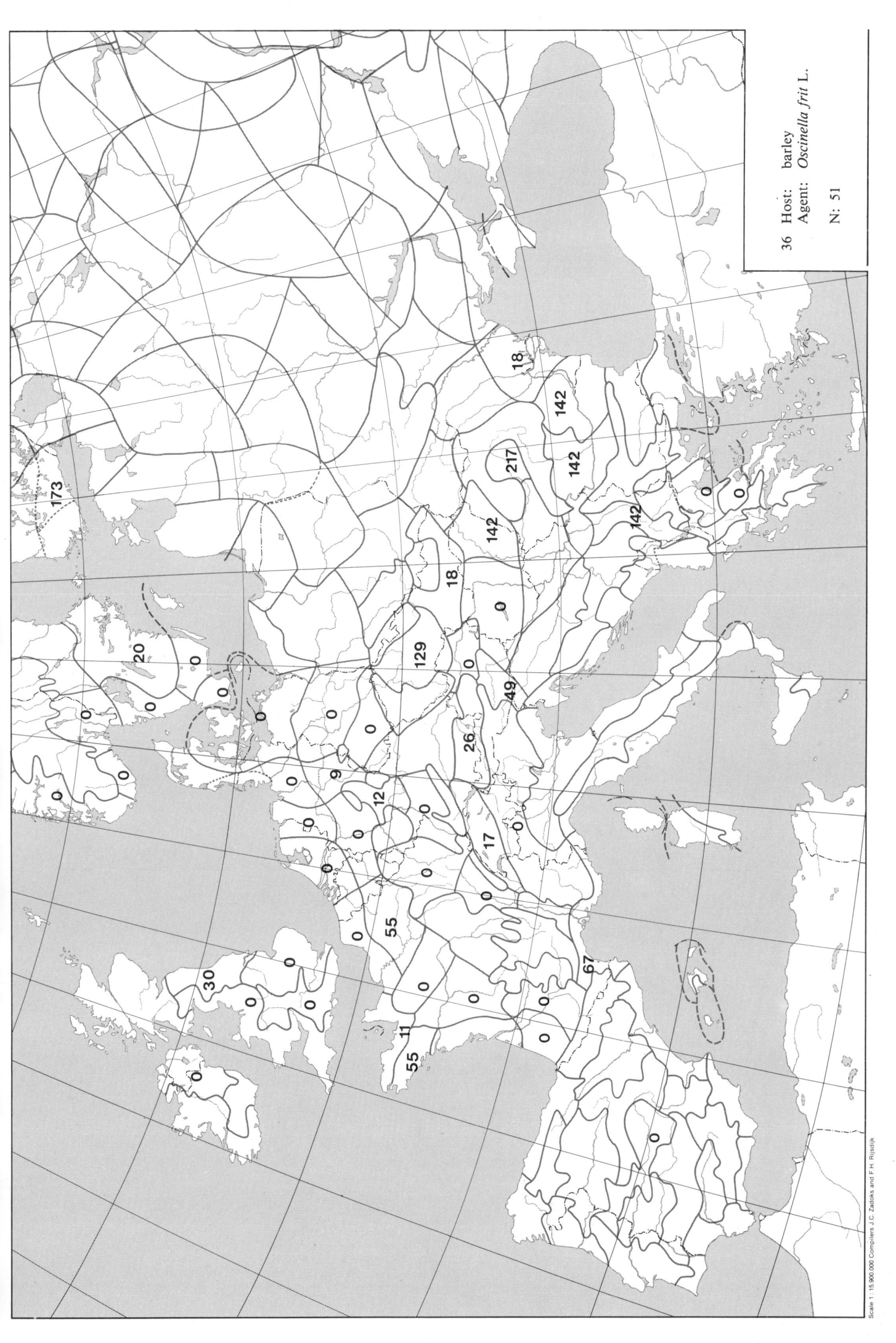

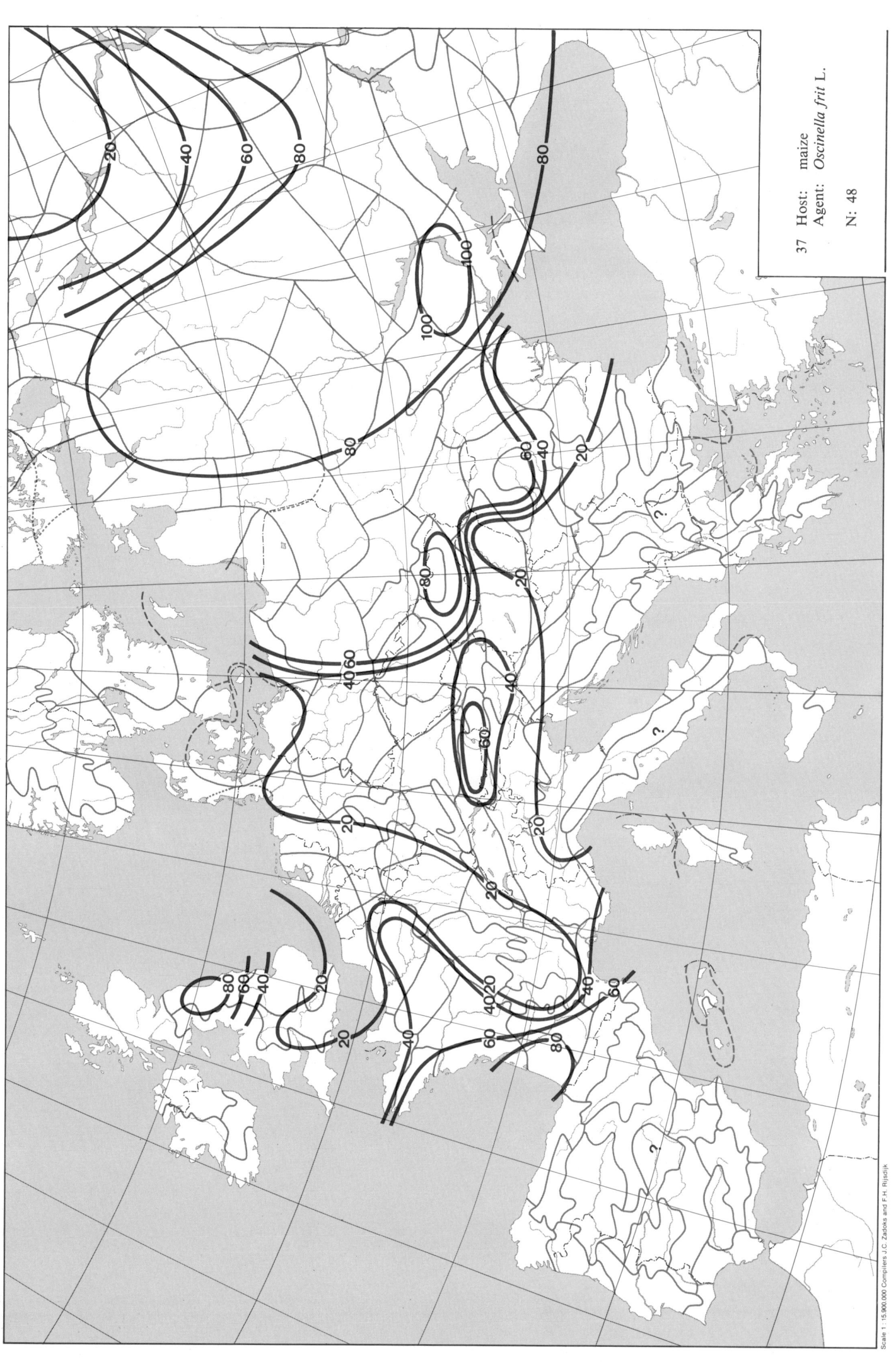

37 Host: maize
Agent: *Oscinella frit* L.

N: 48

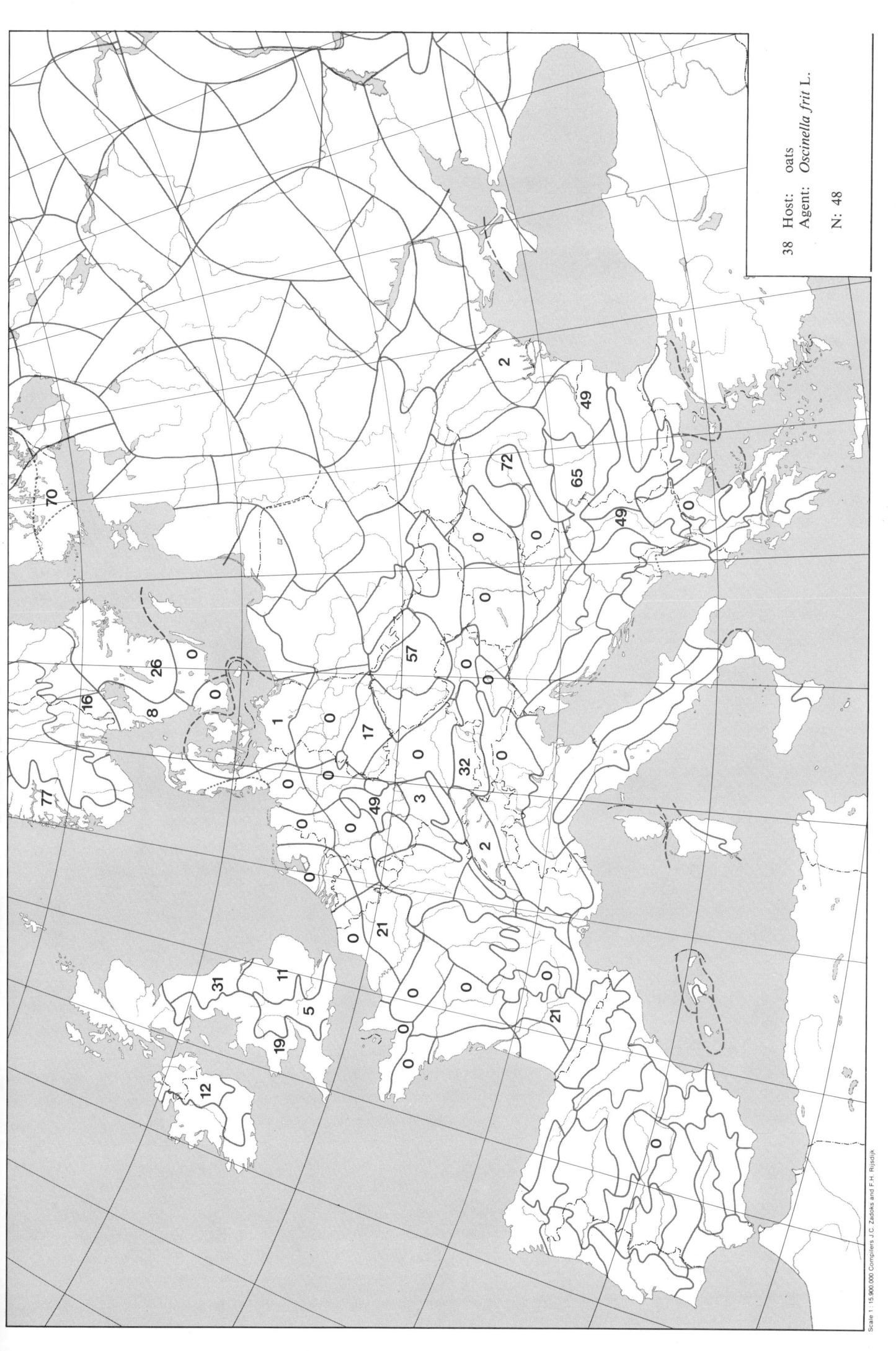

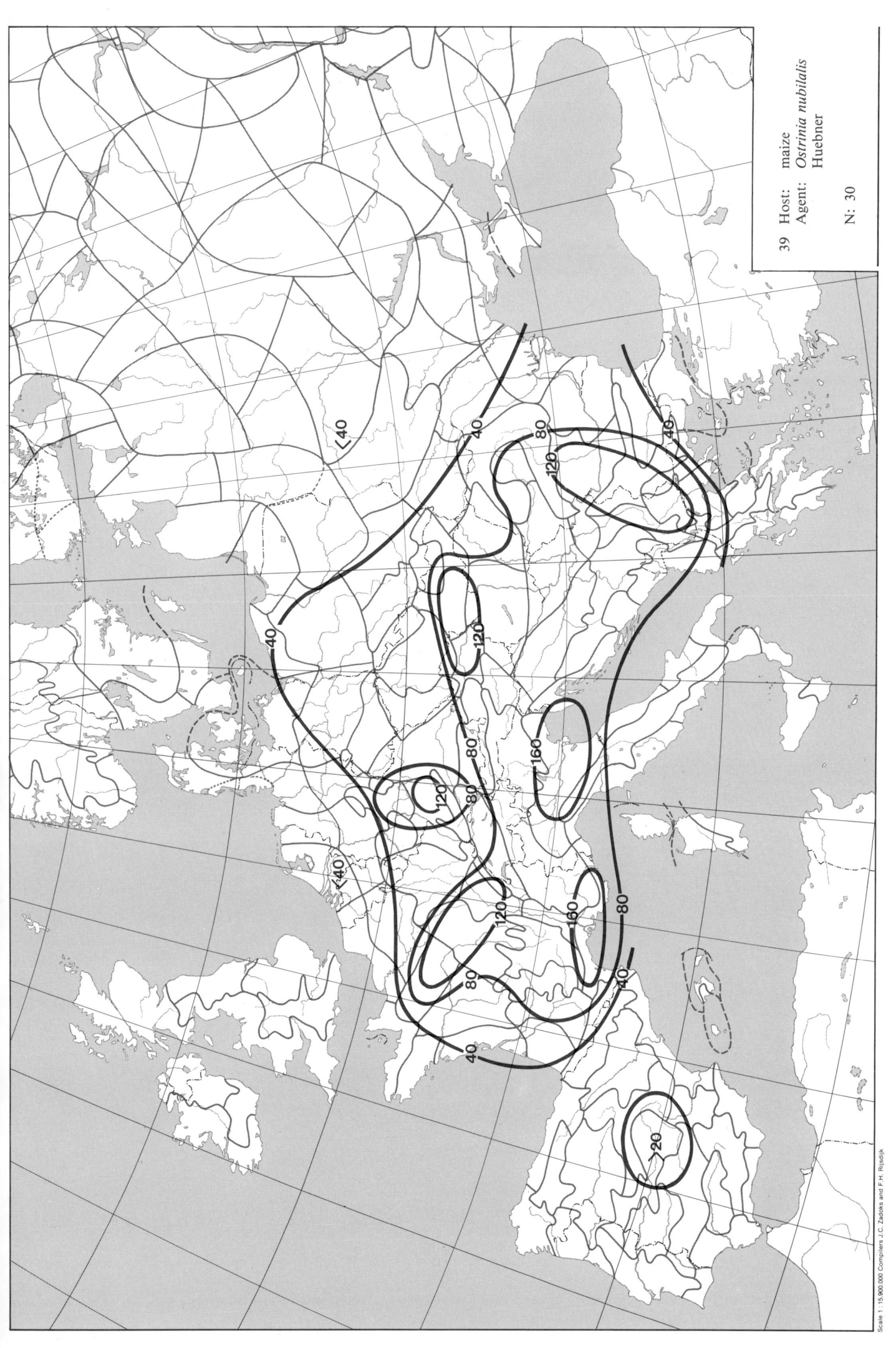

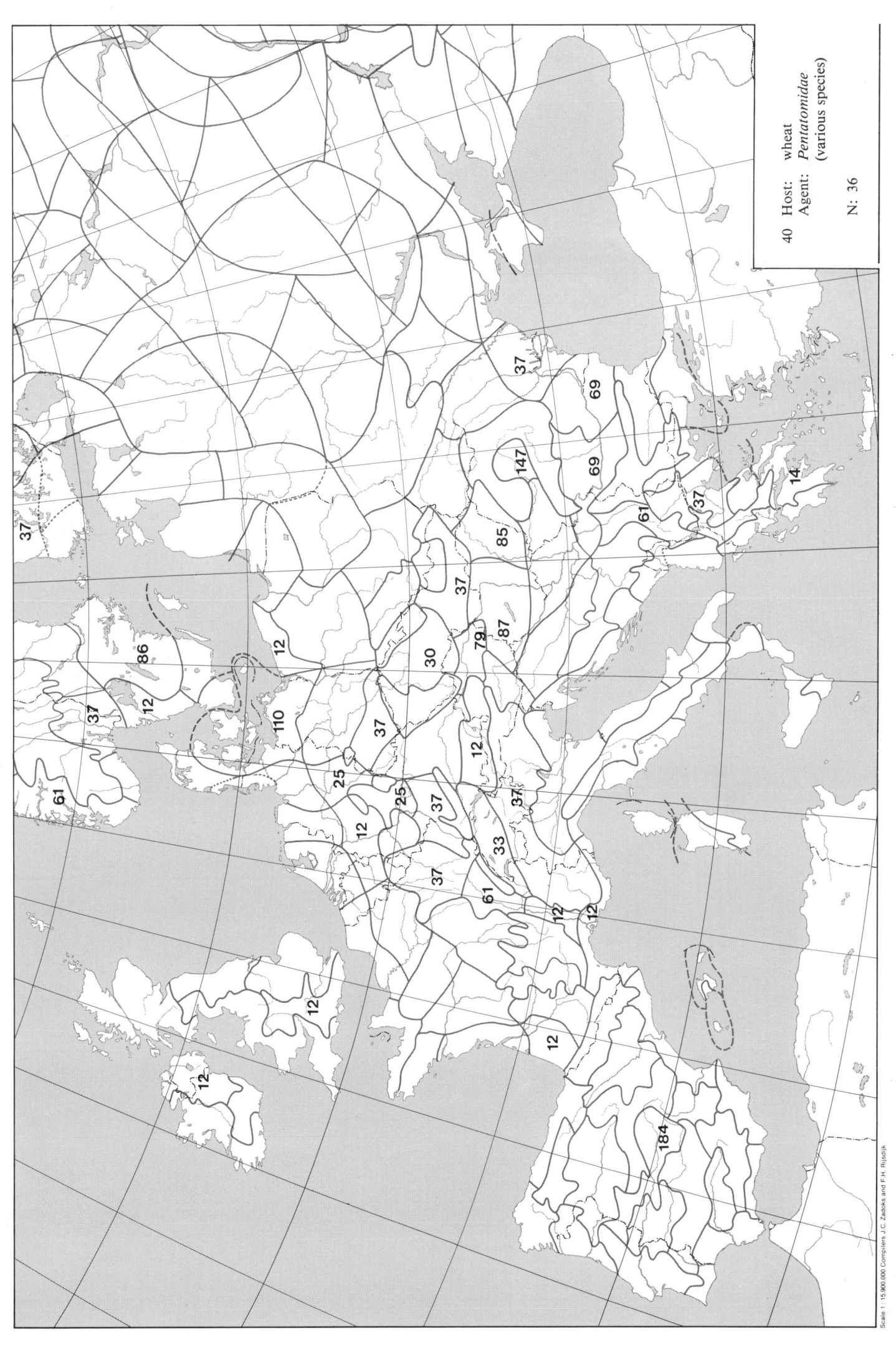

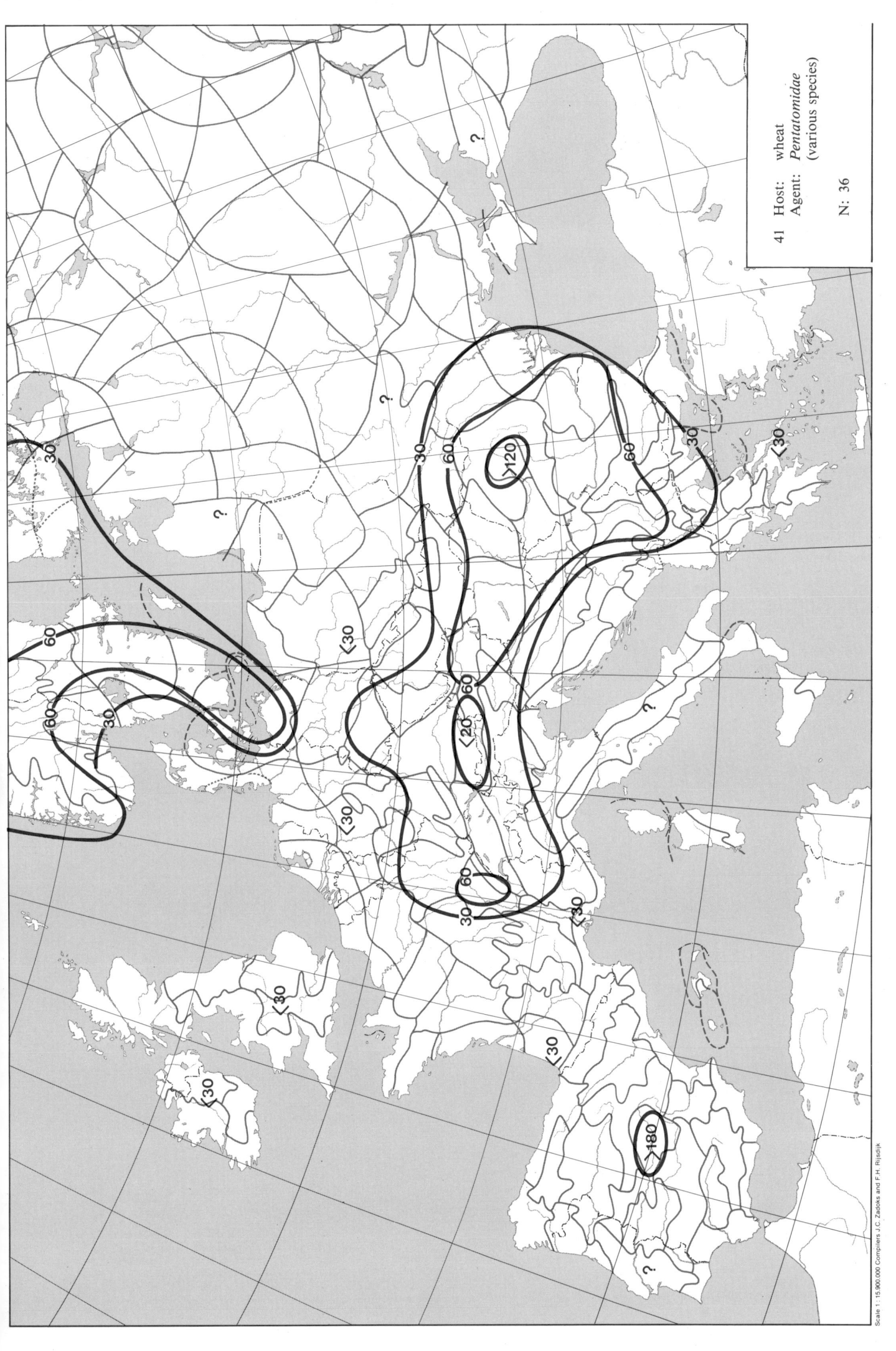

41 Host: wheat
Agent: *Pentatomidae* (various species)

N: 36

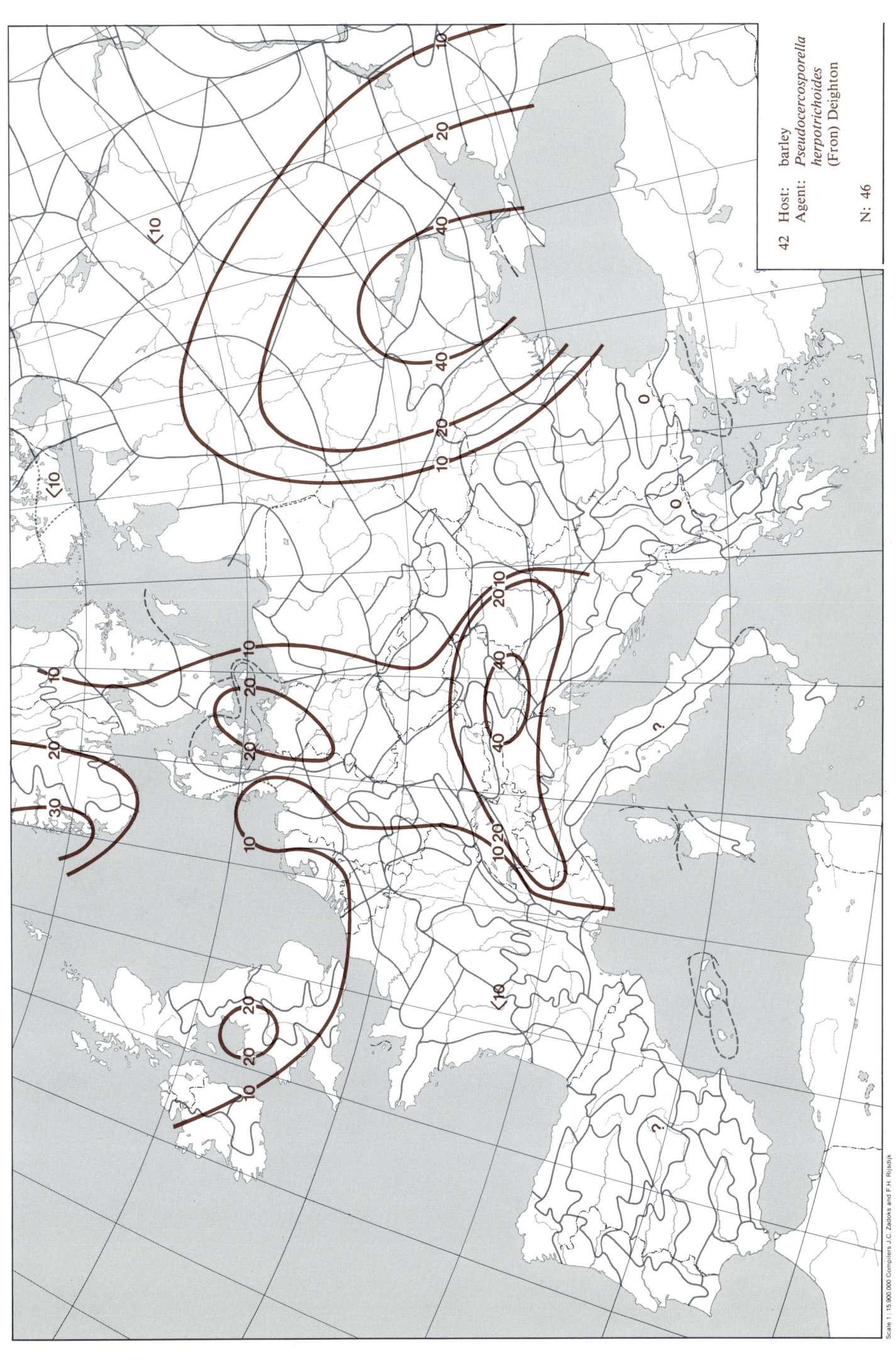

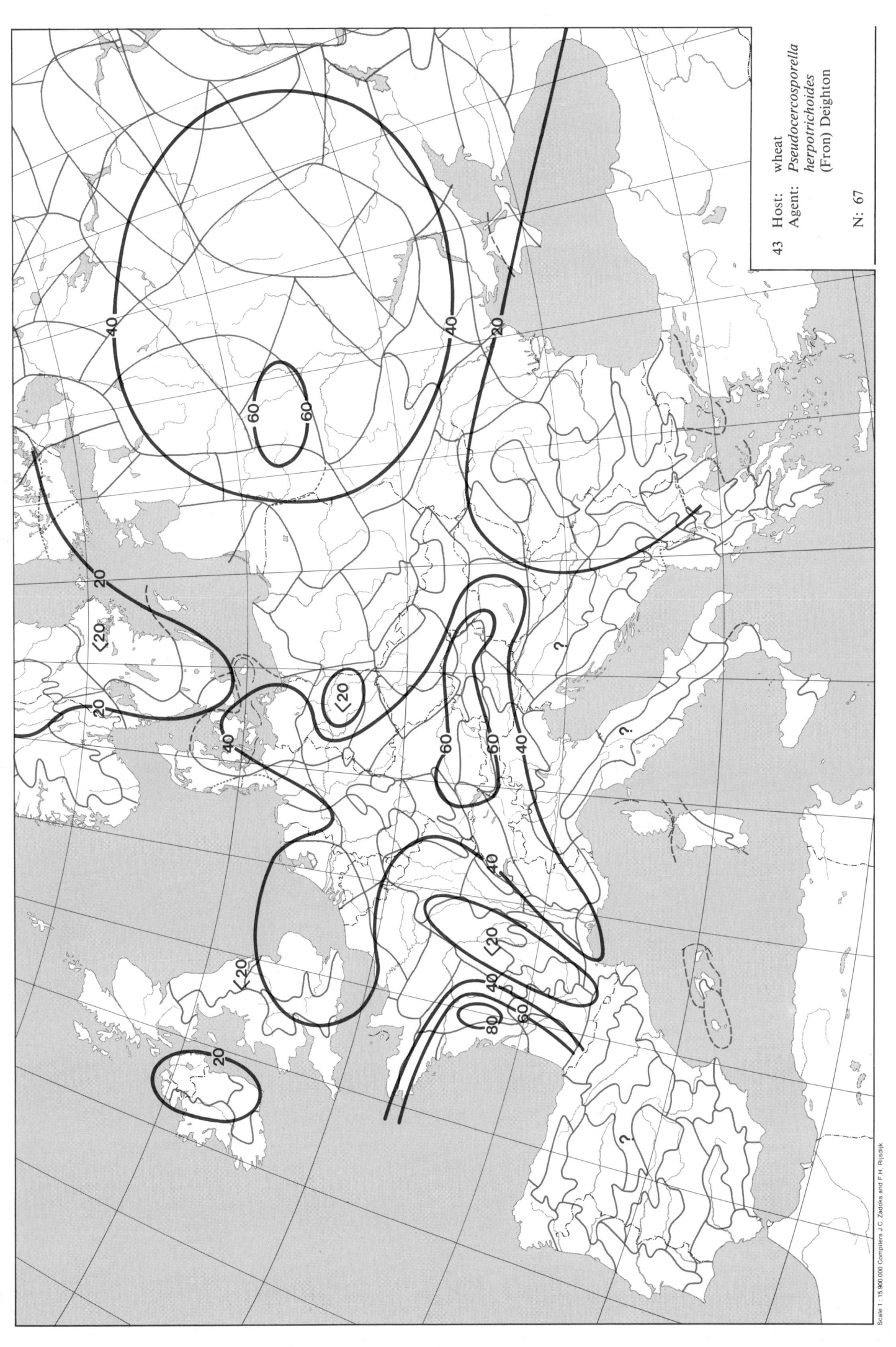

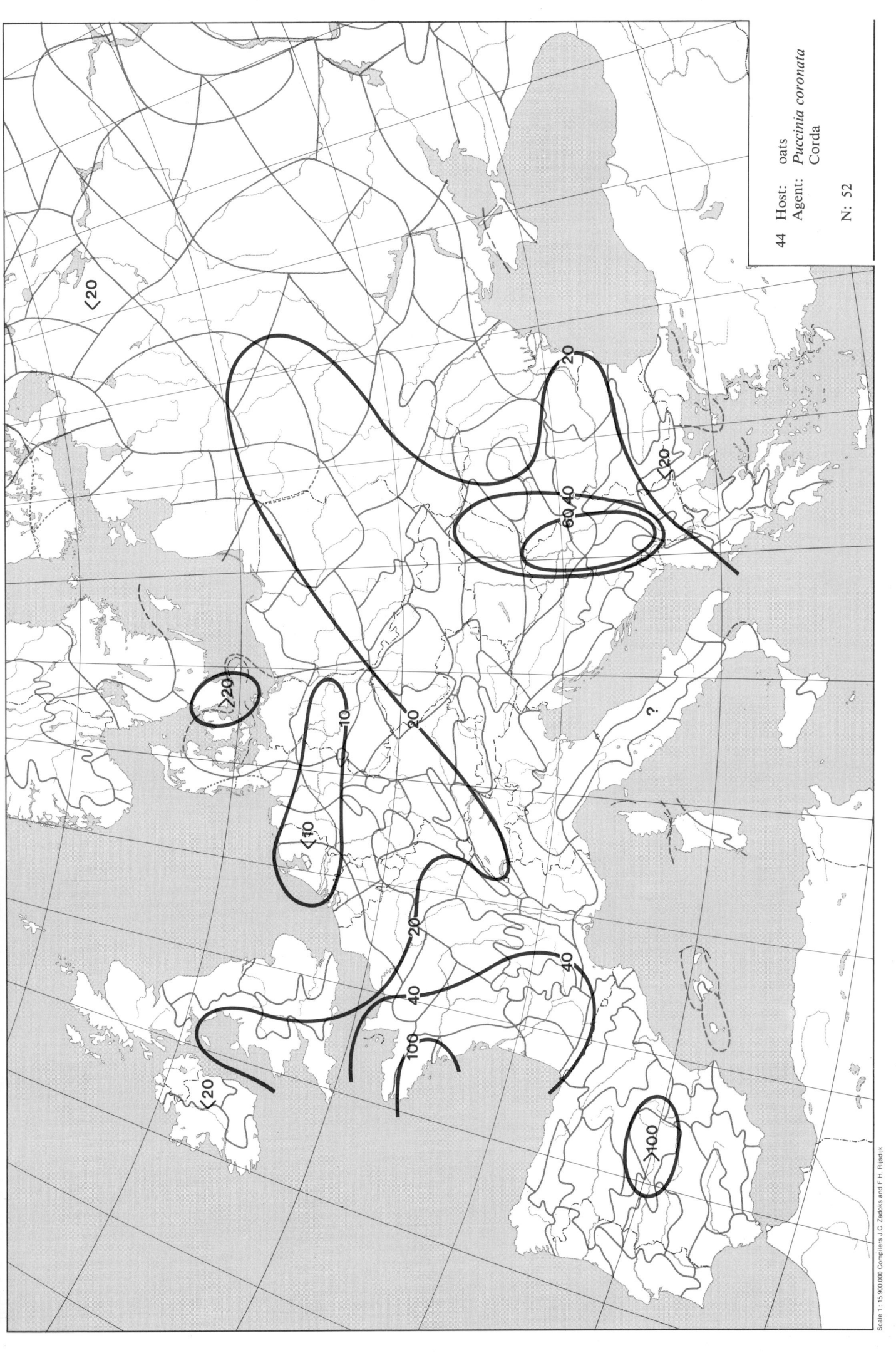

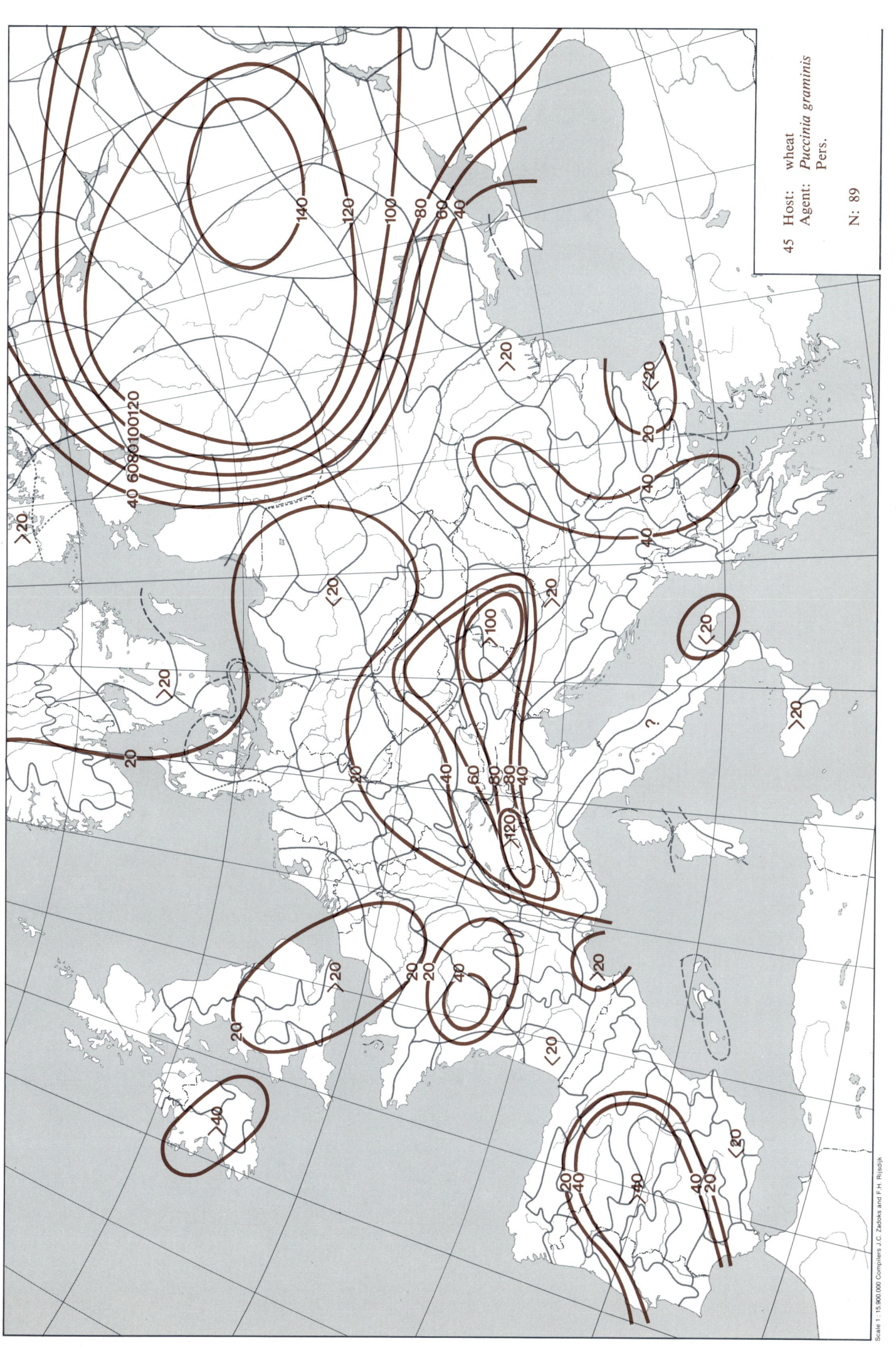

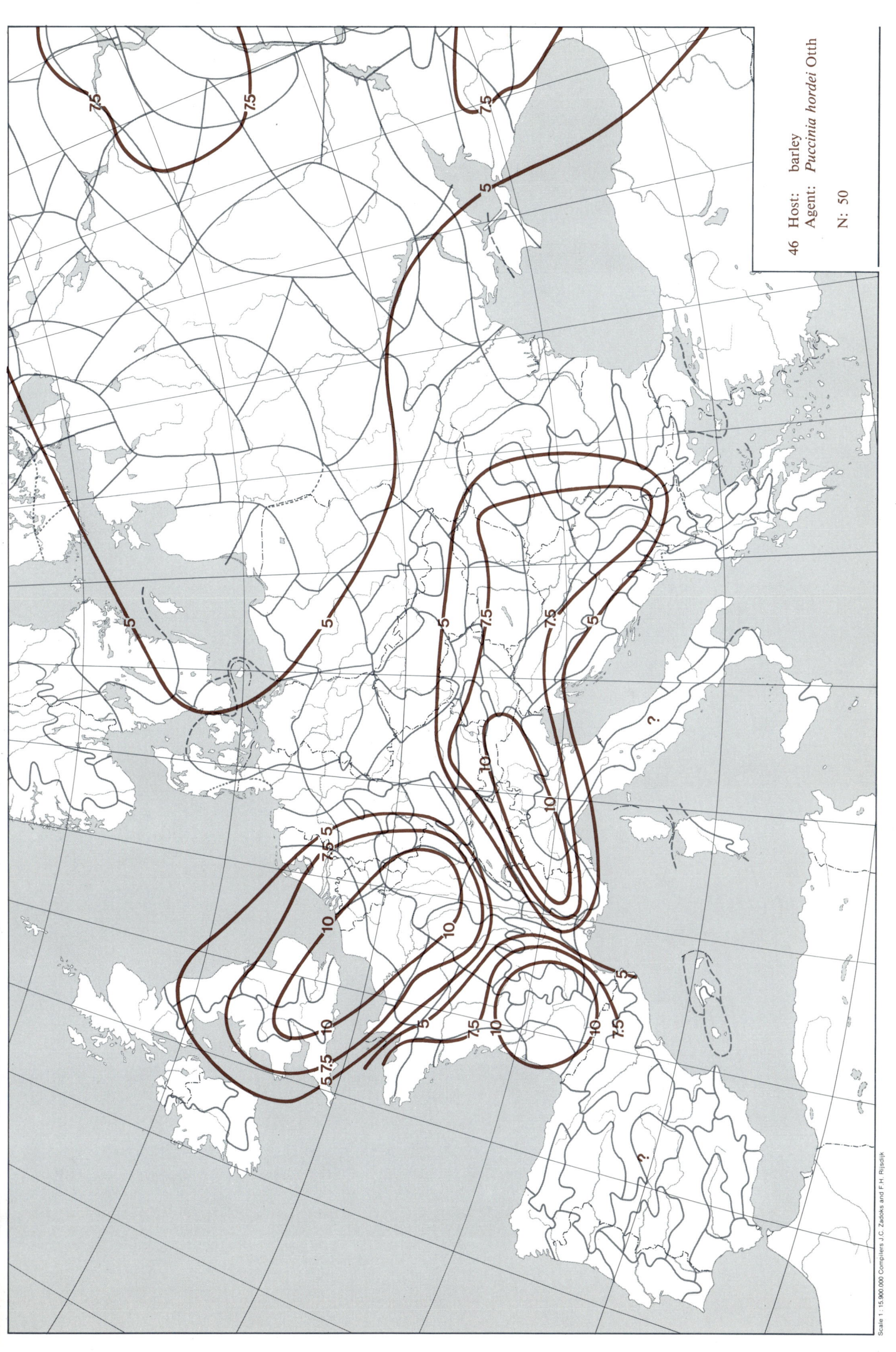

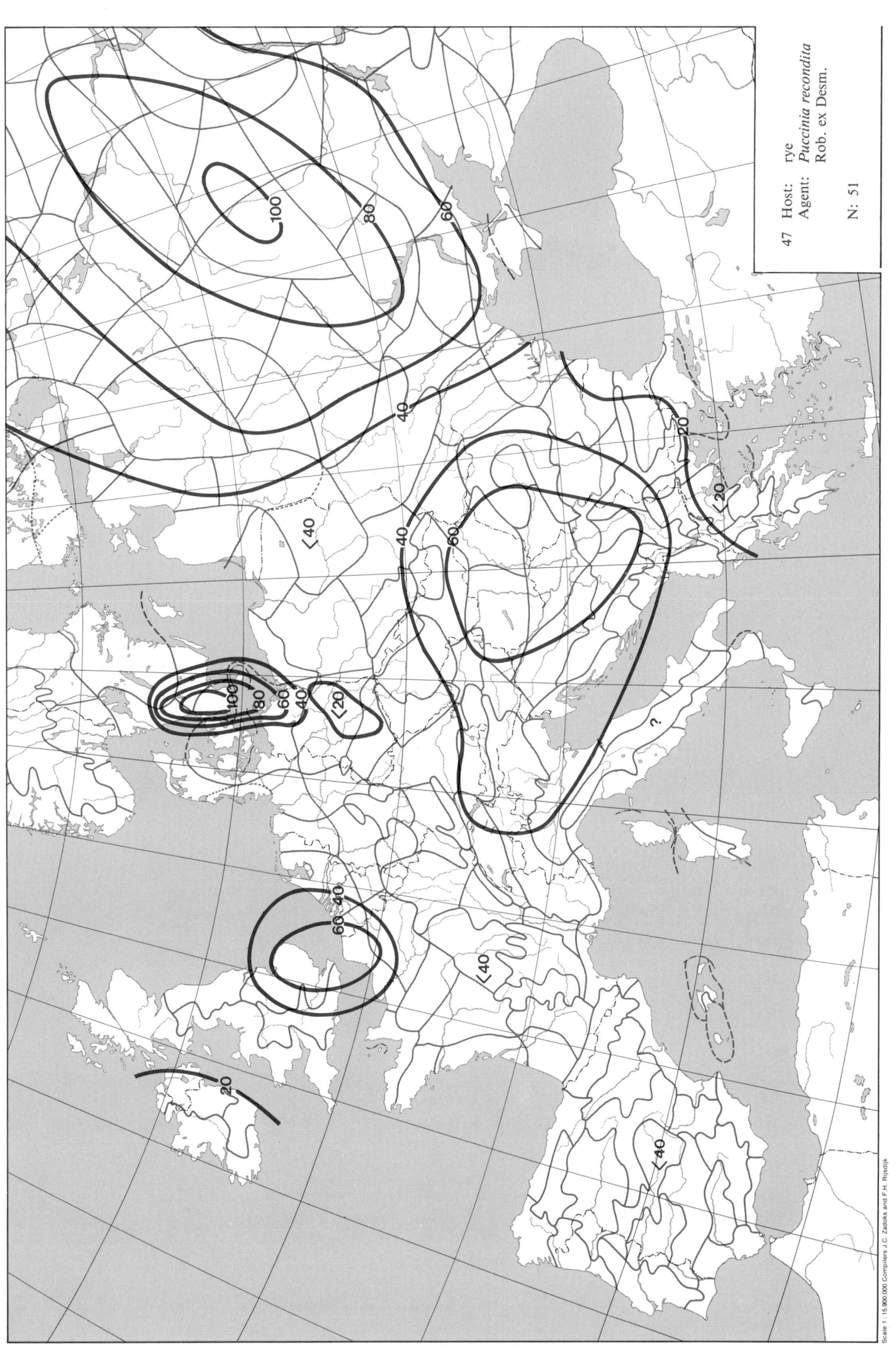

47 Host: rye
Agent: *Puccinia recondita* Rob. ex Desm.

N: 51

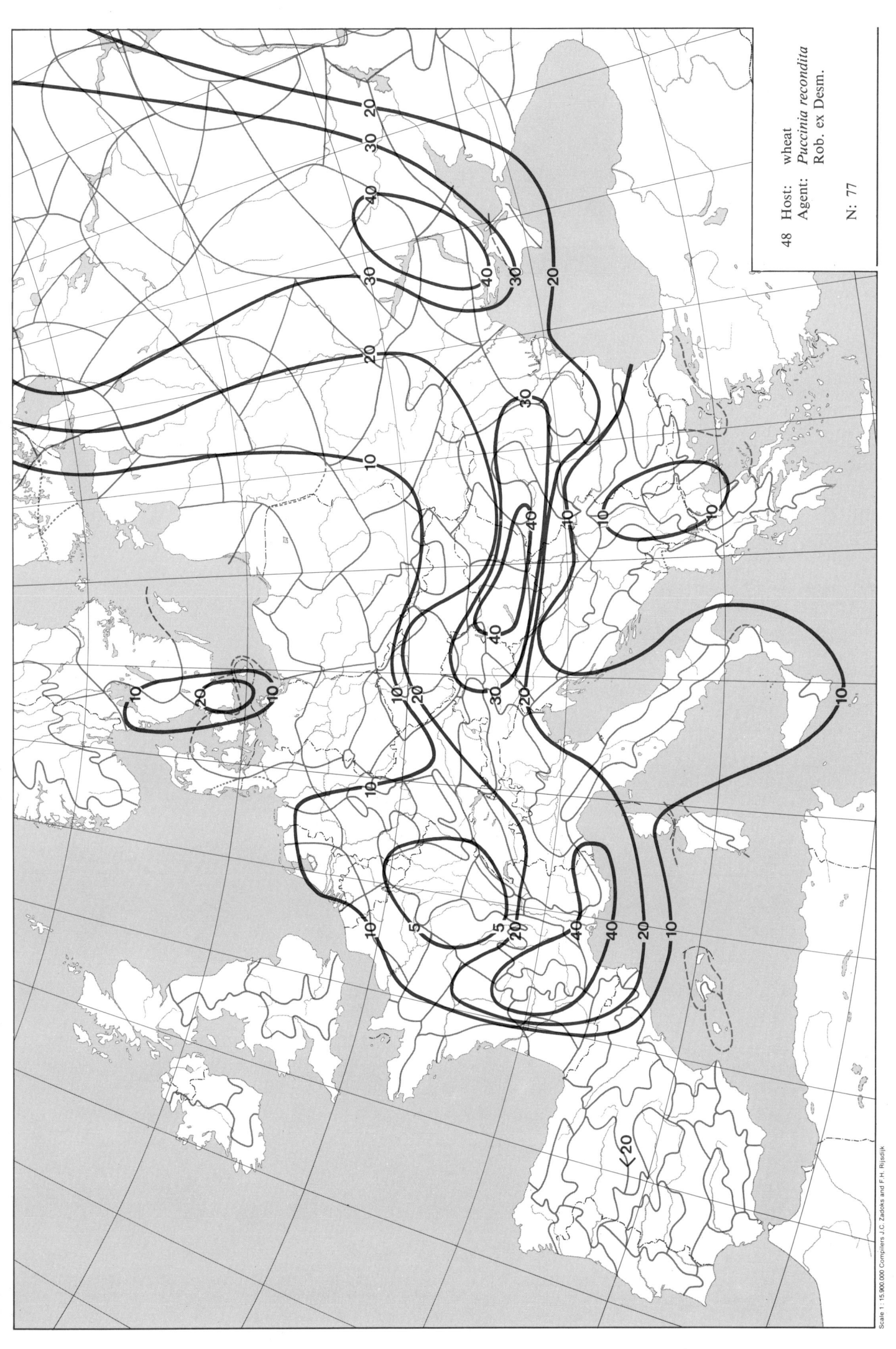

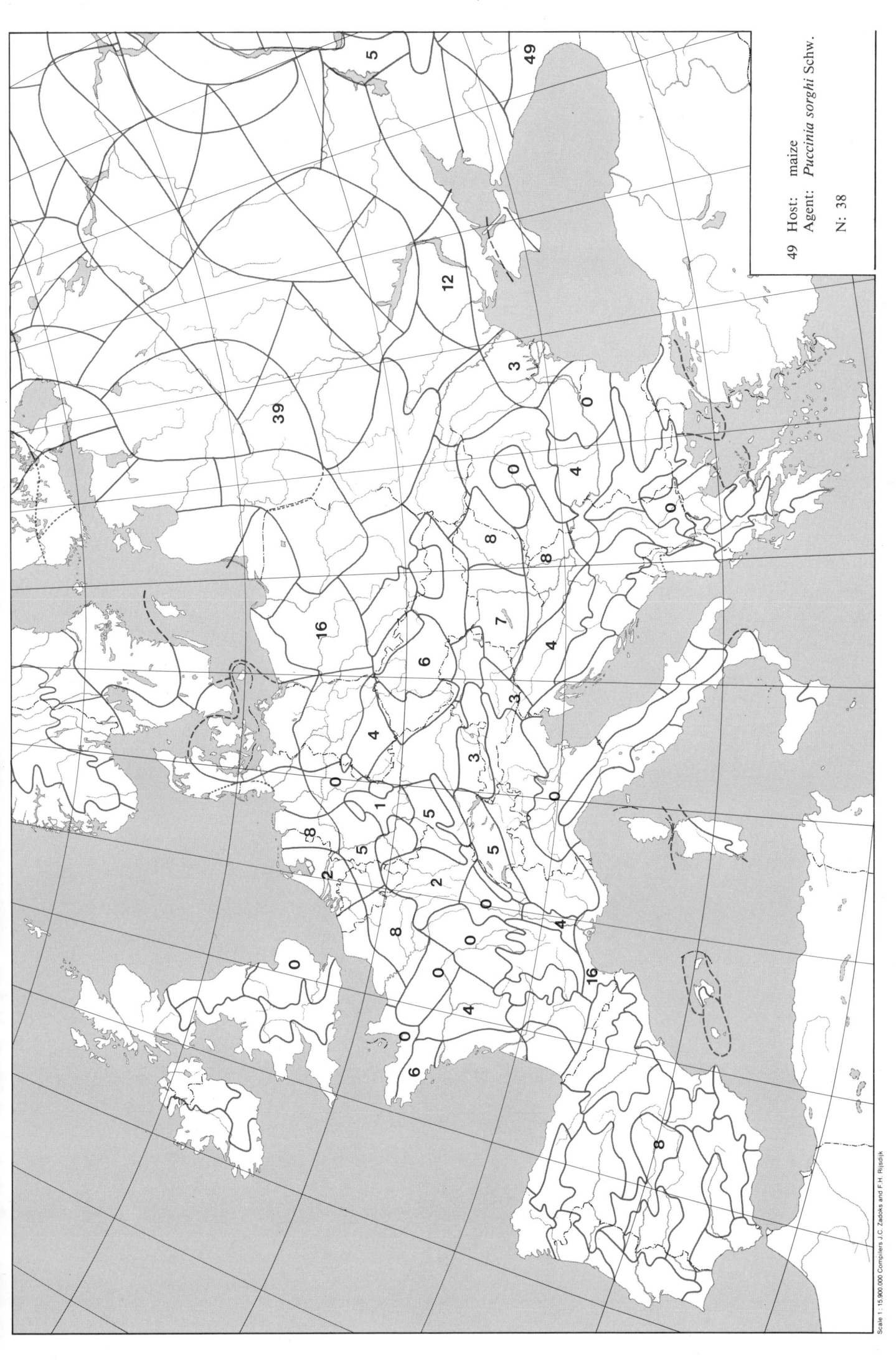

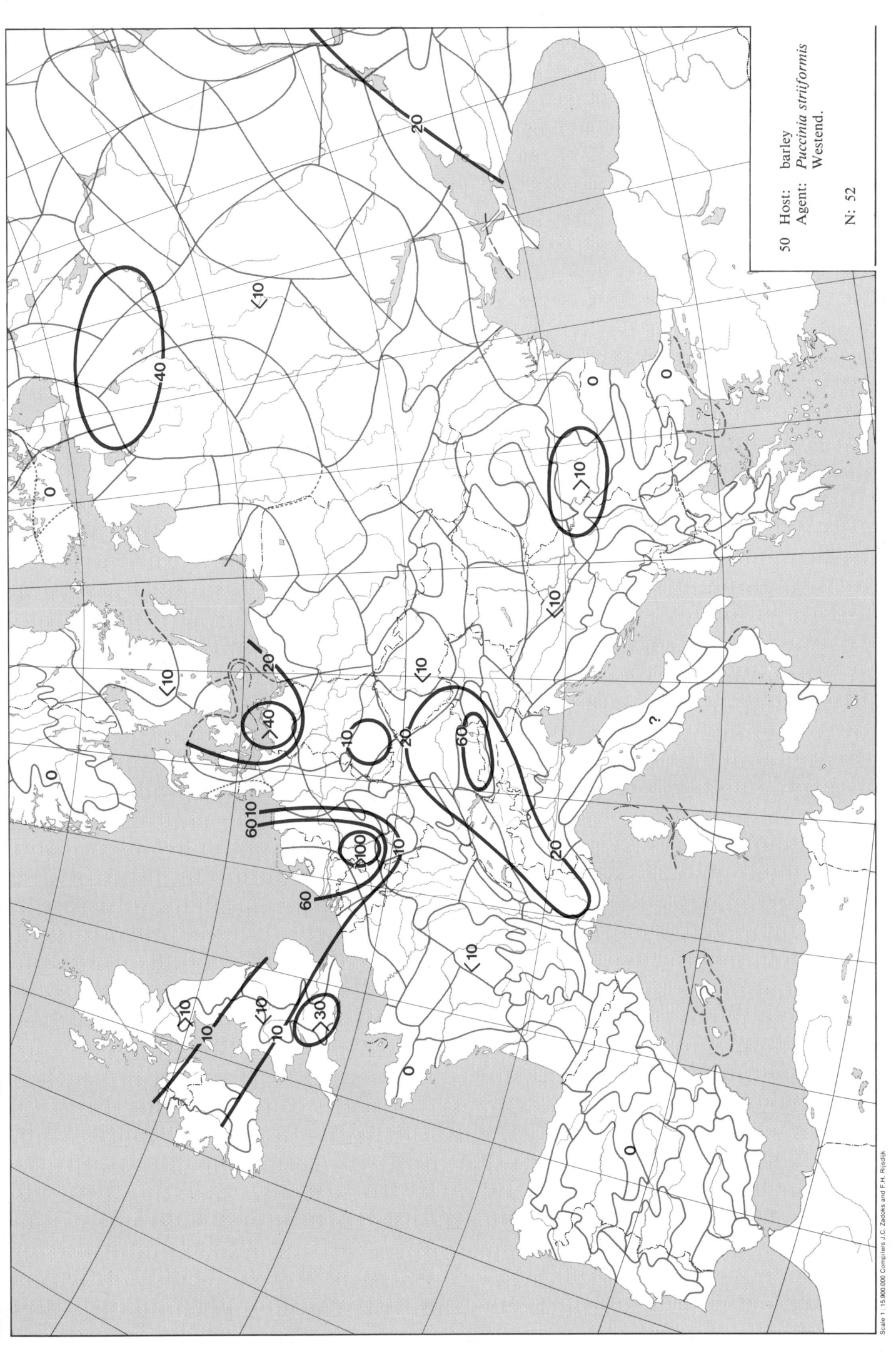

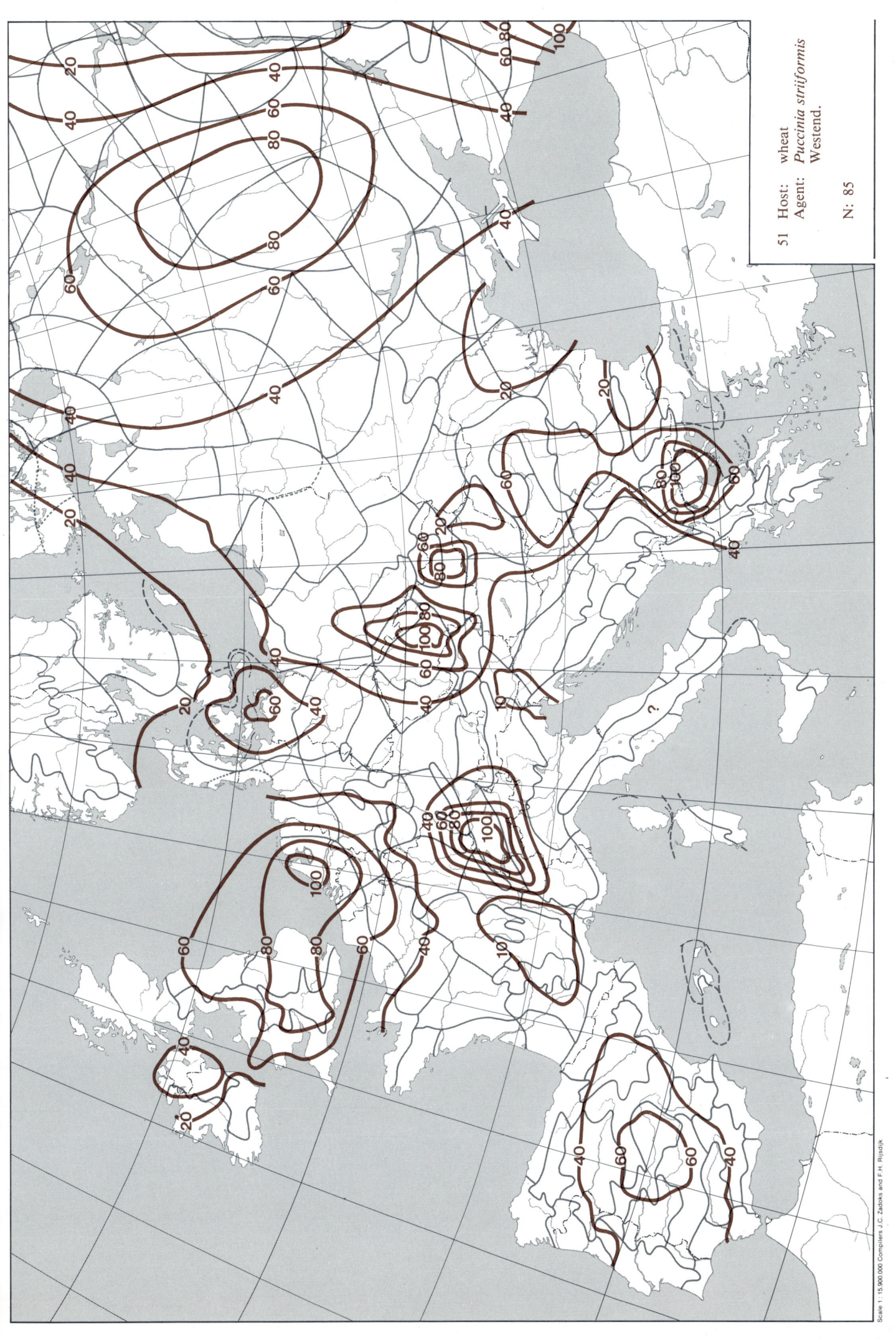

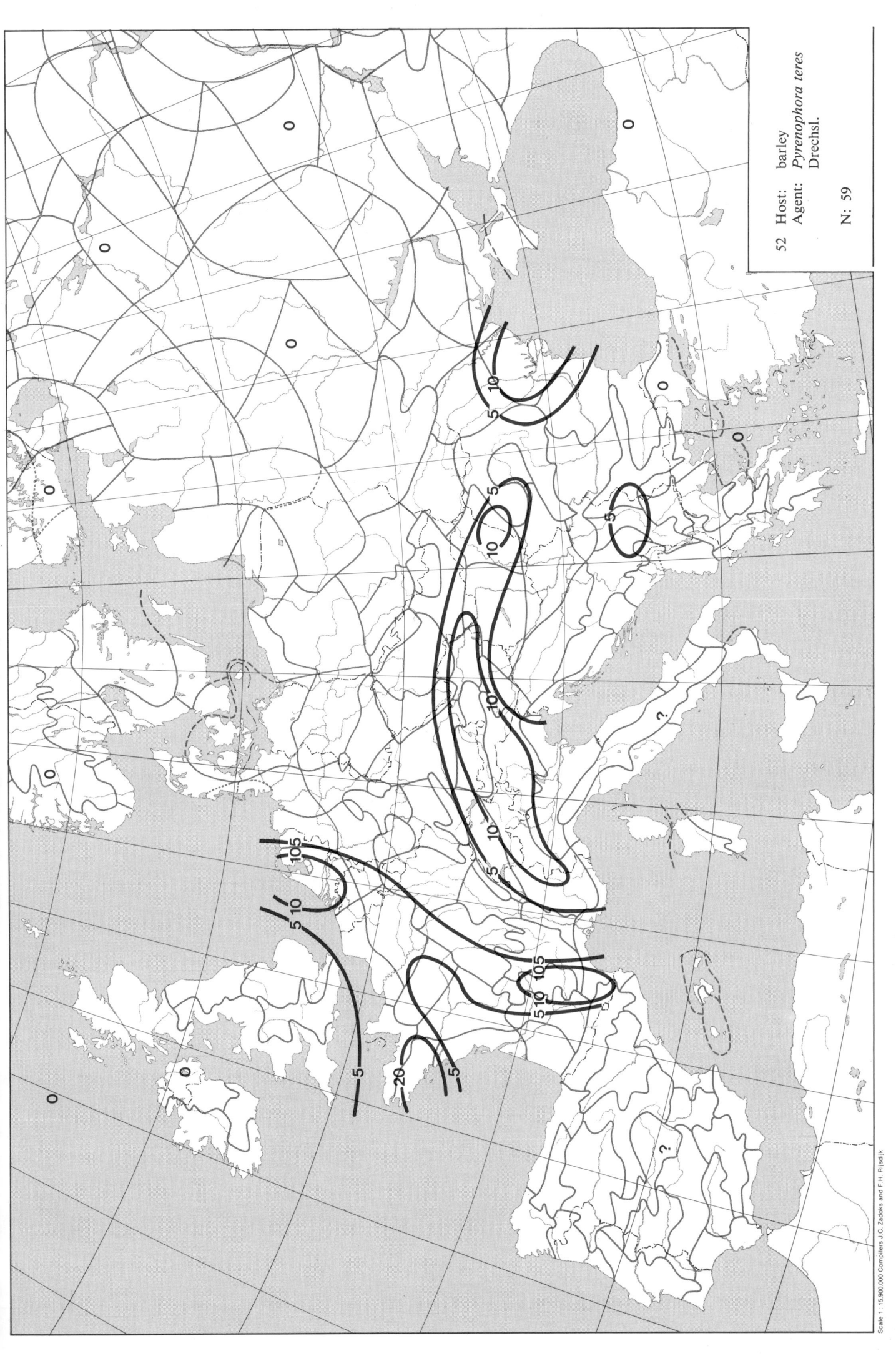

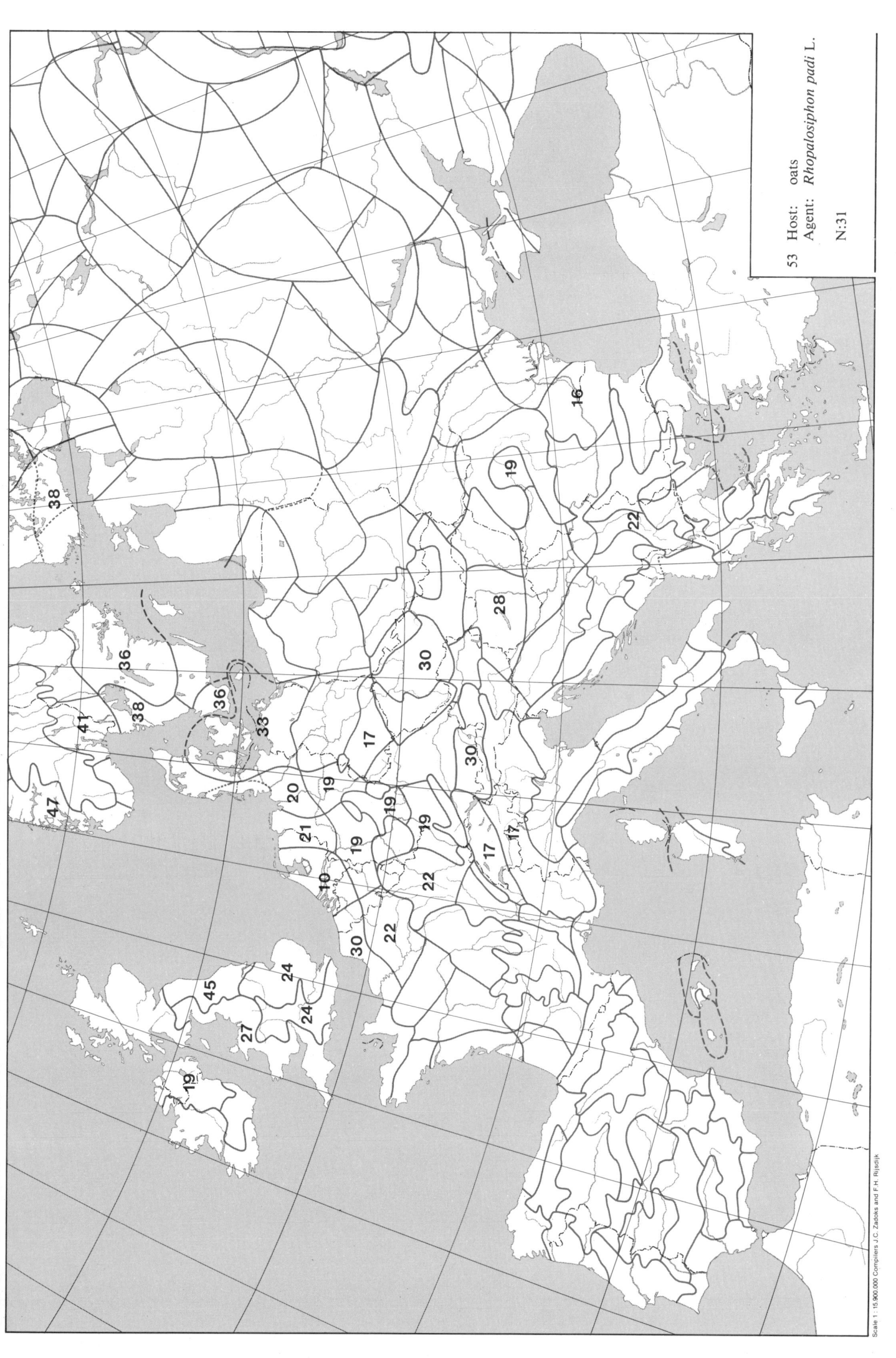

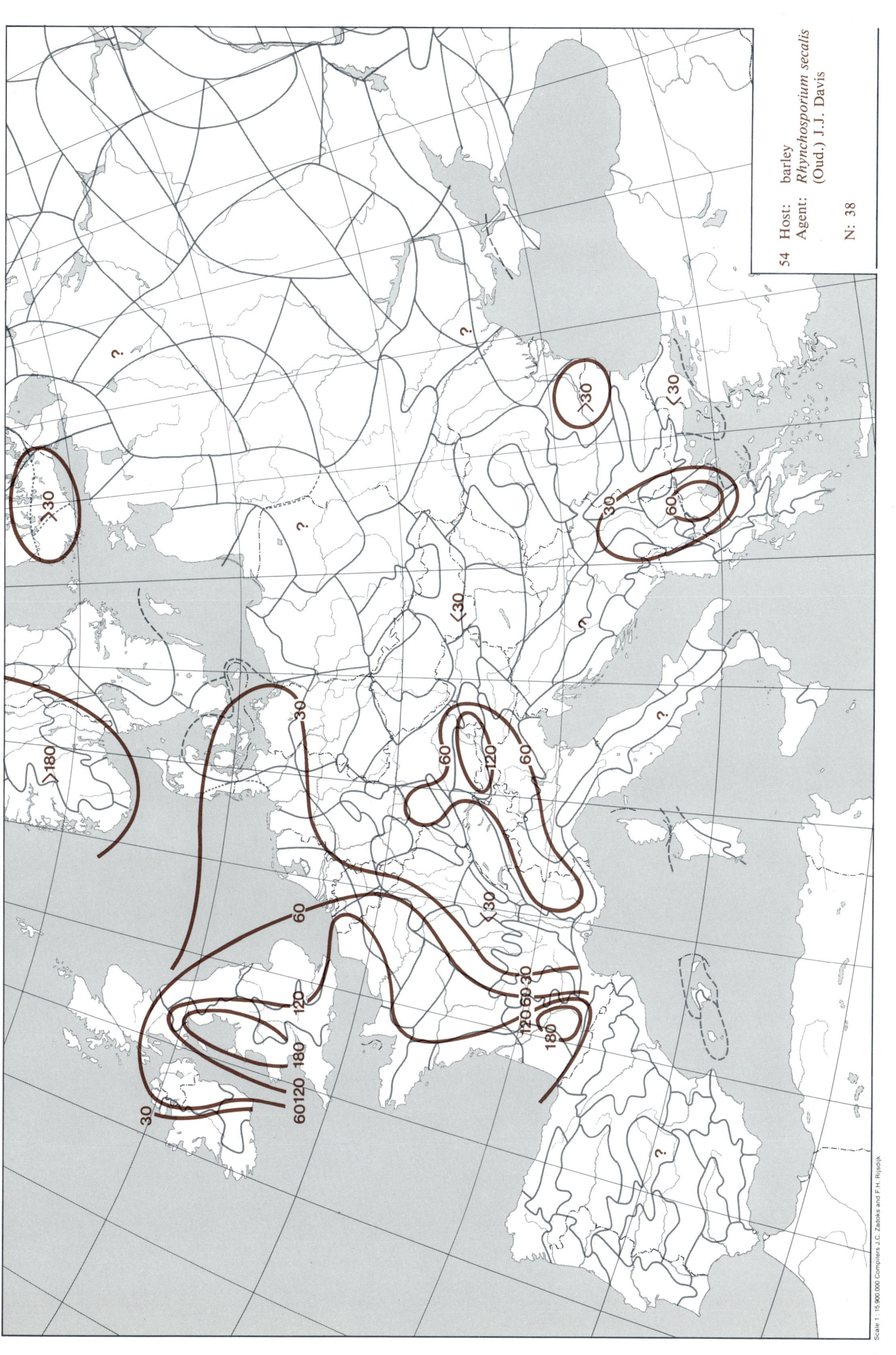

54 Host: barley
Agent: *Rhynchosporium secalis* (Oud.) J.J. Davis

N: 38

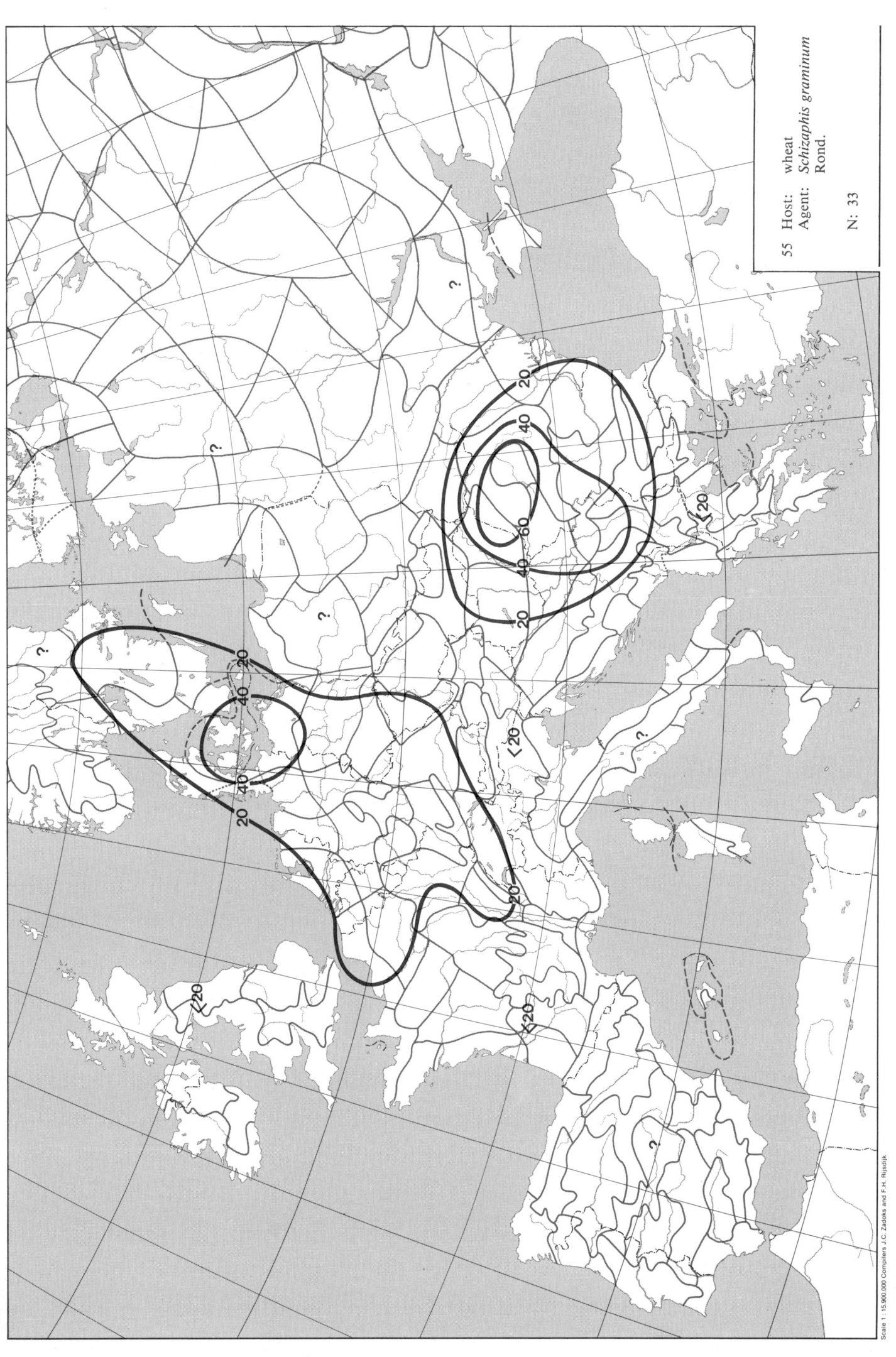

55 Host: wheat
Agent: *Schizaphis graminum* Rond.

N: 33

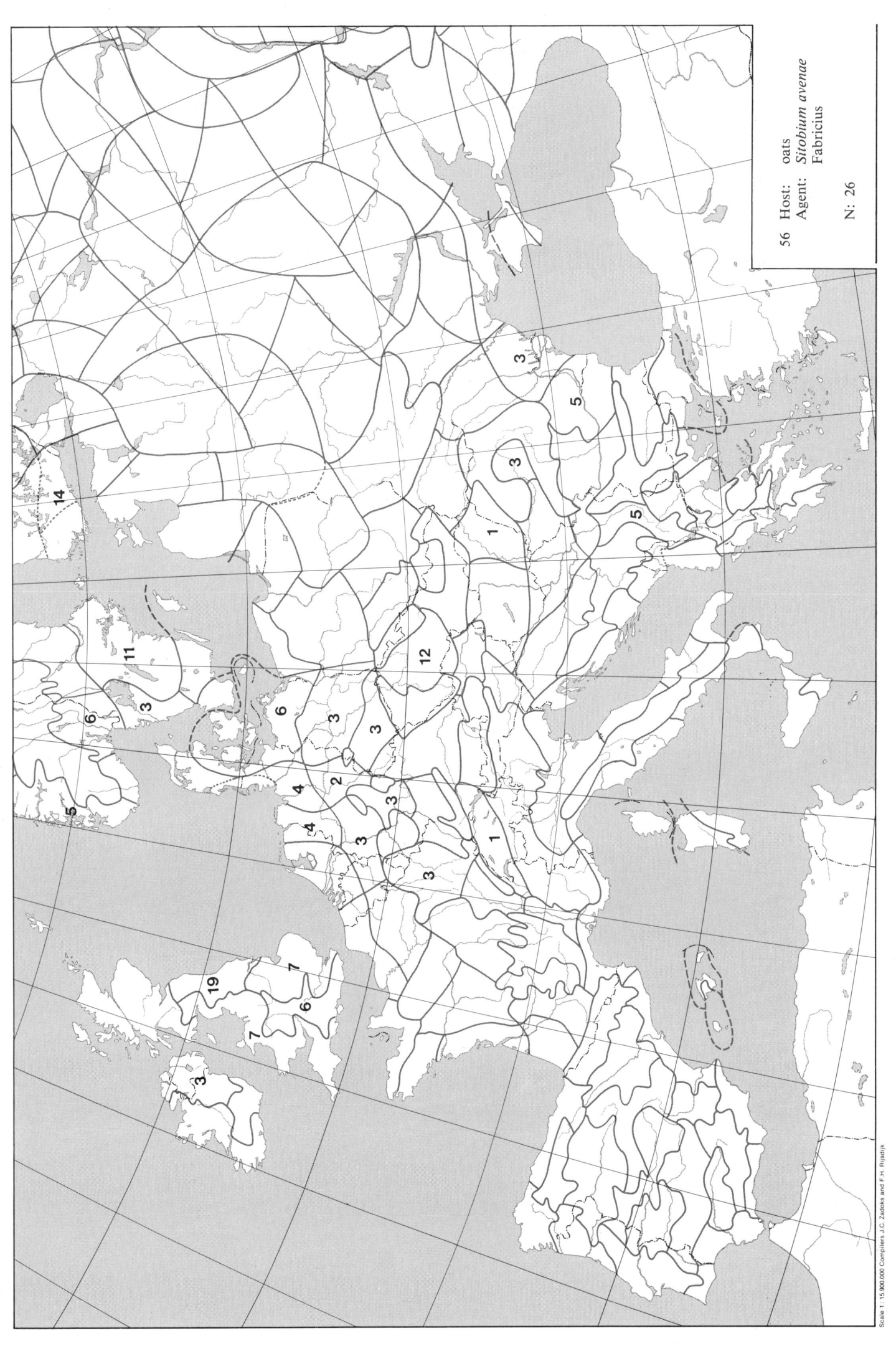

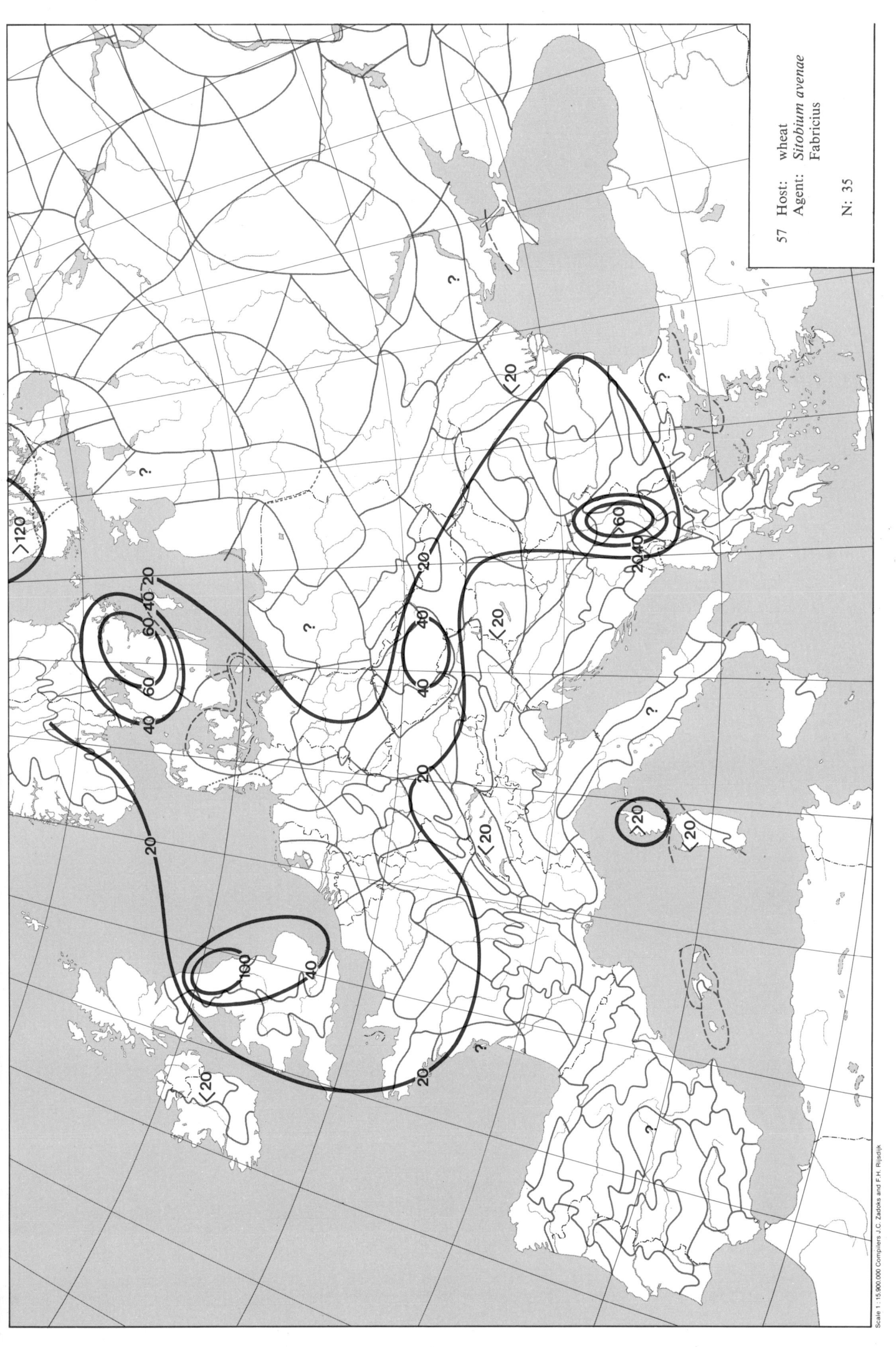

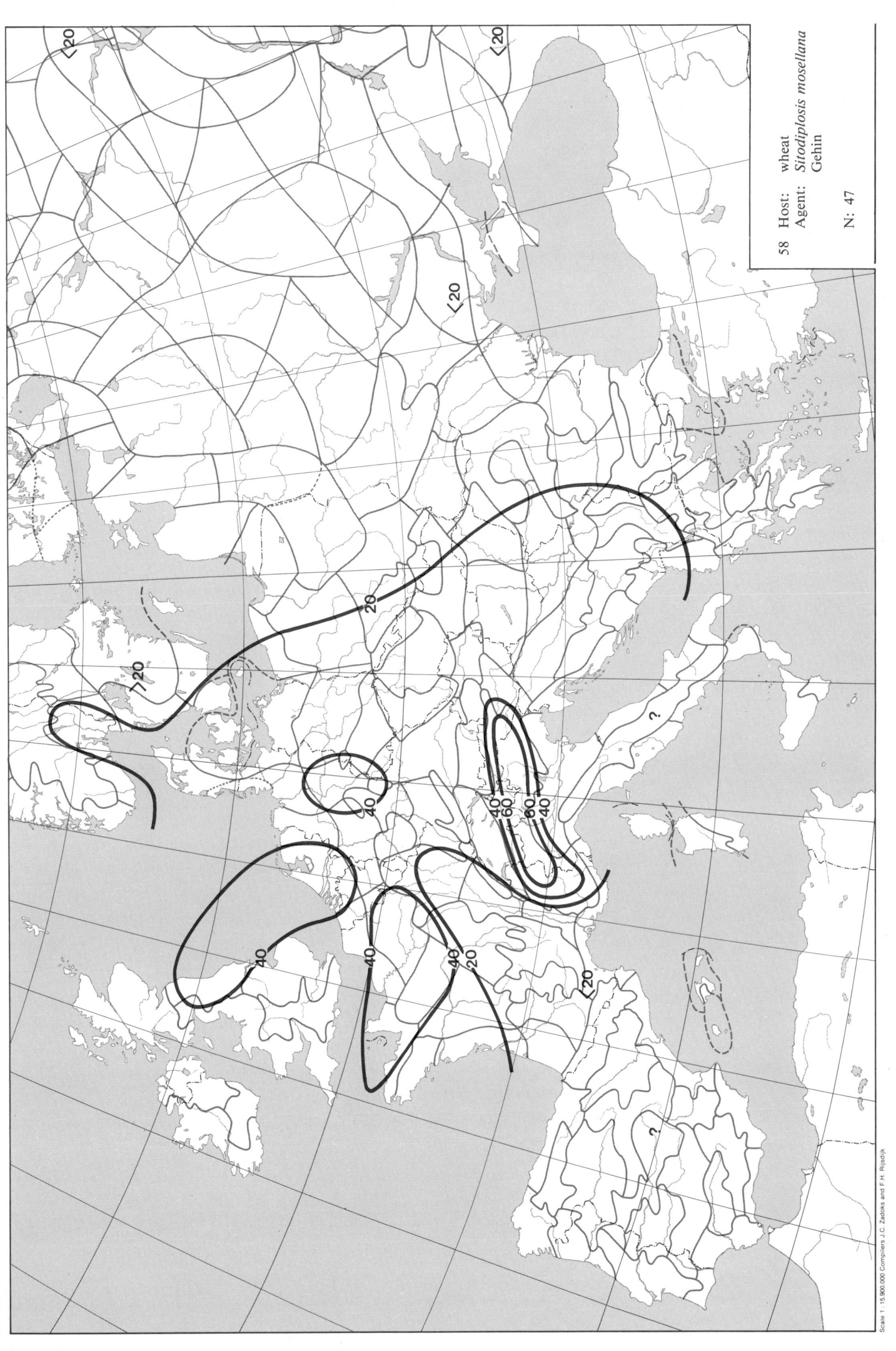

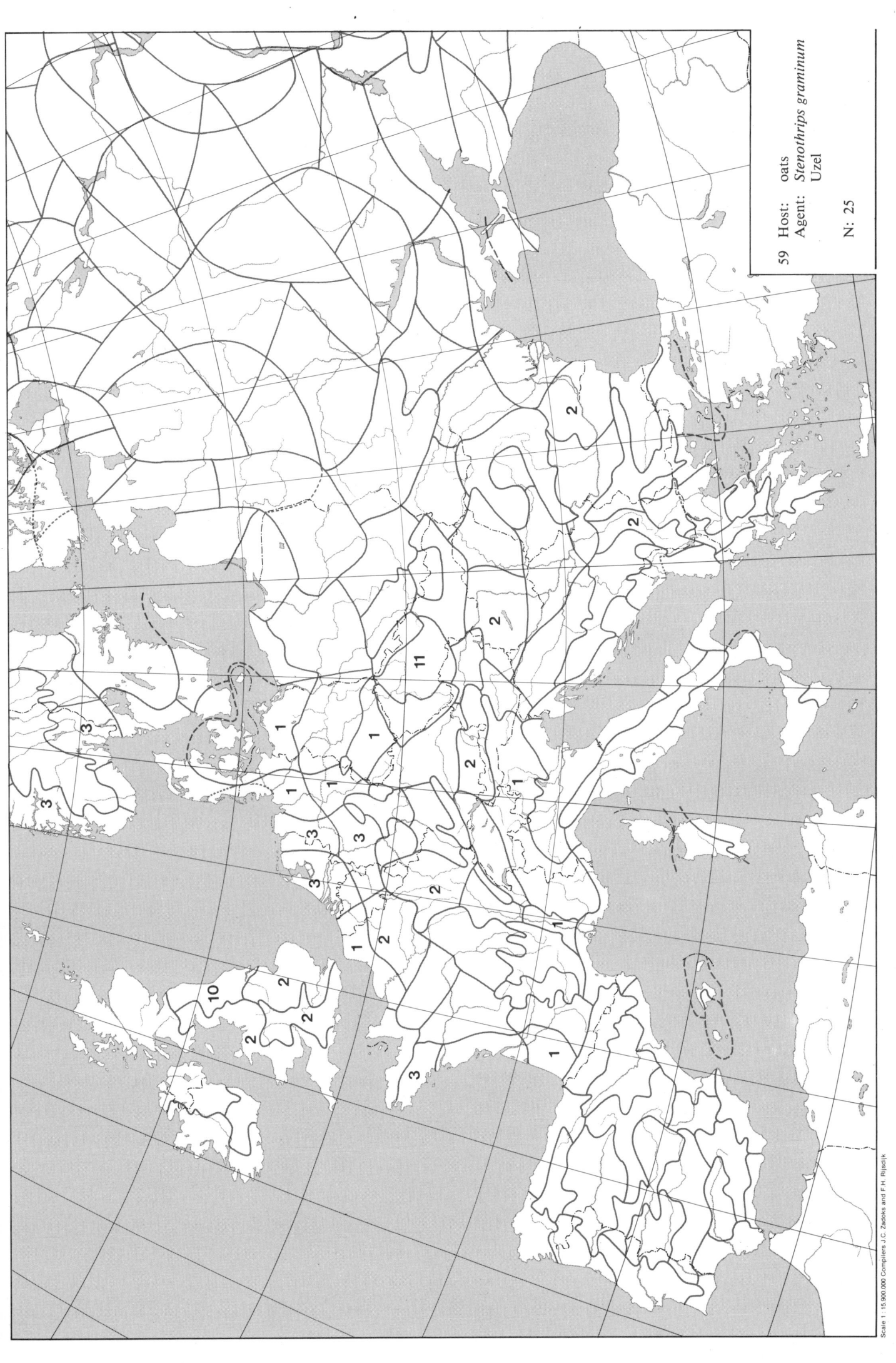

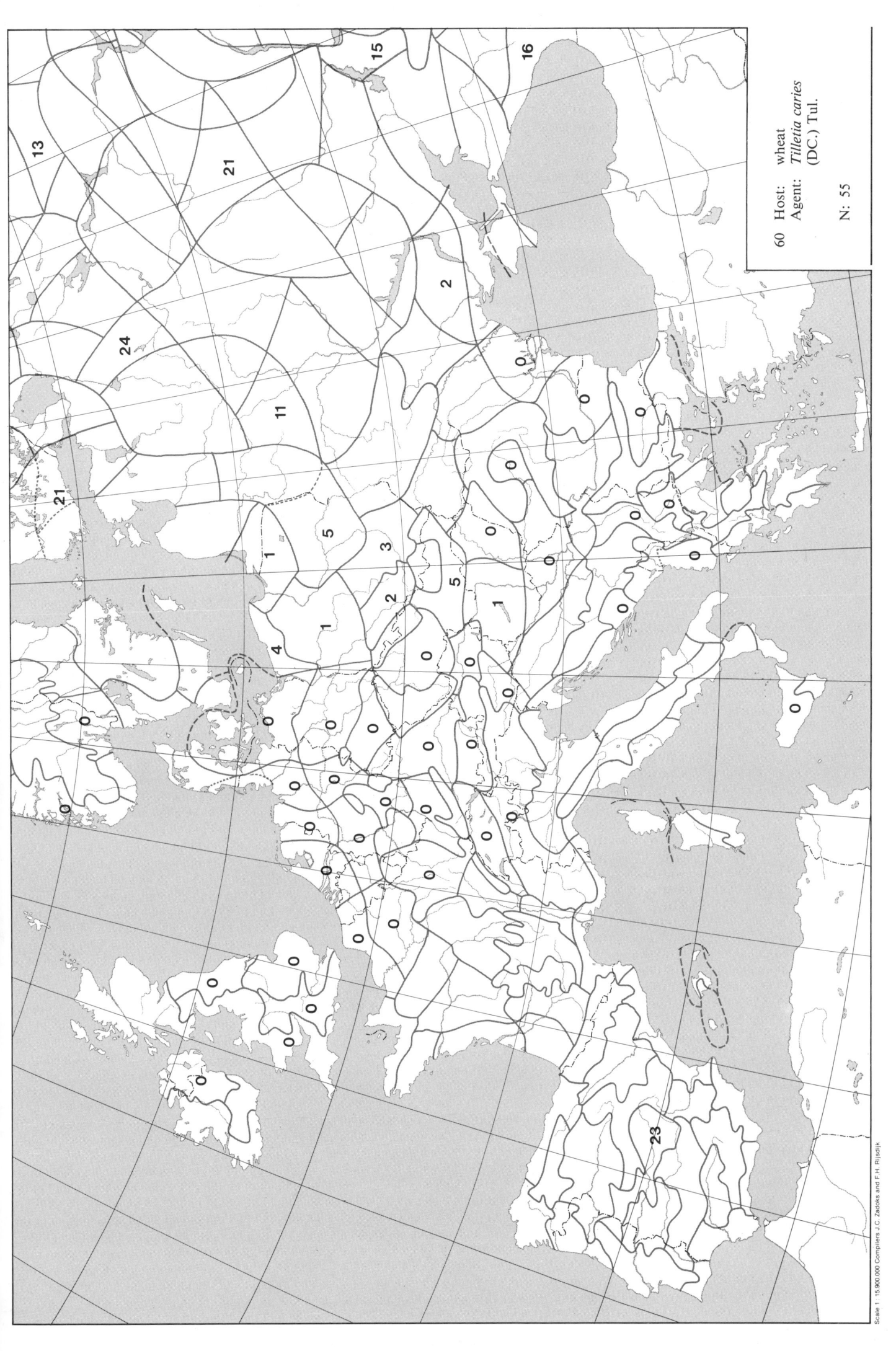

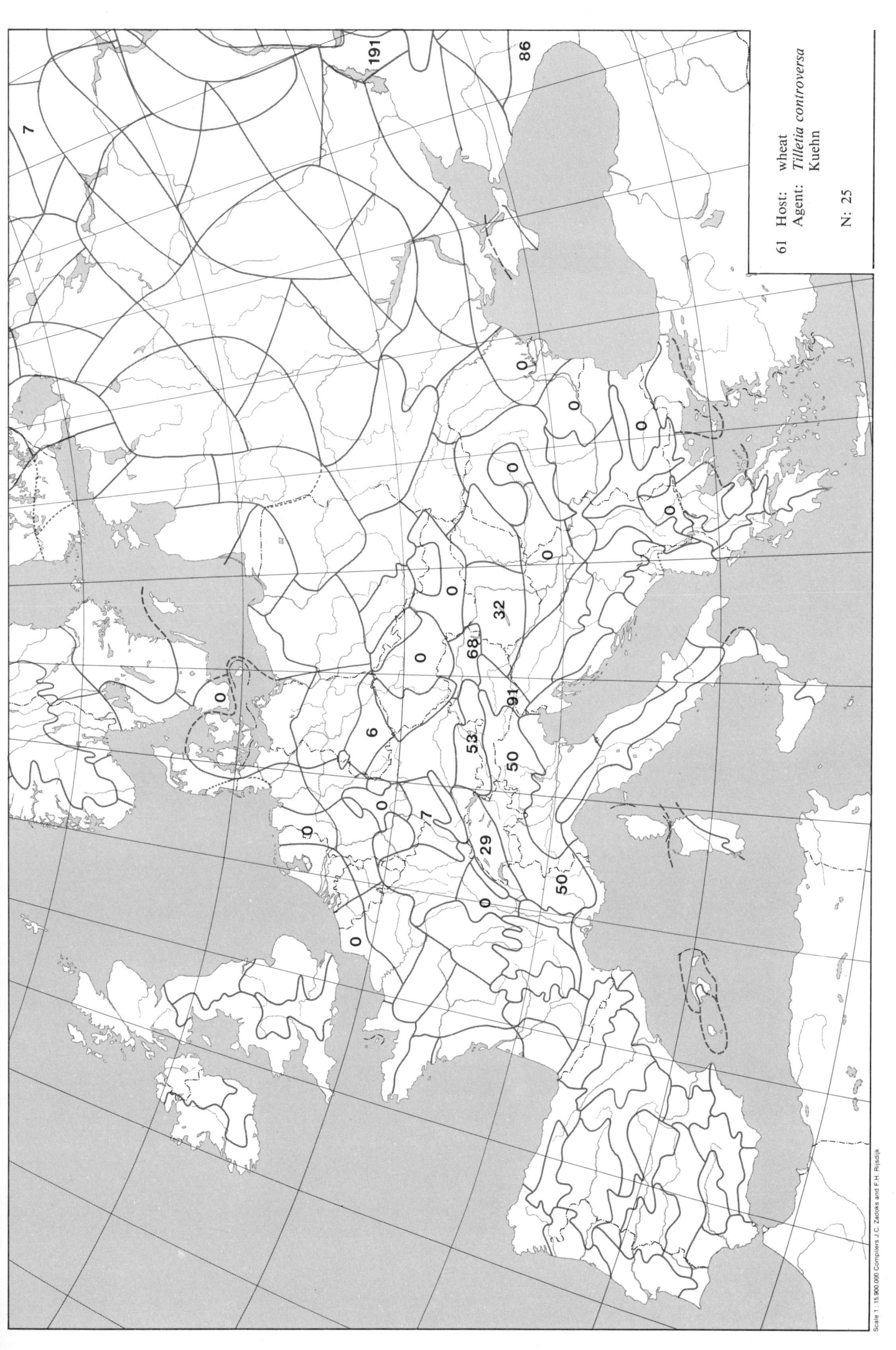

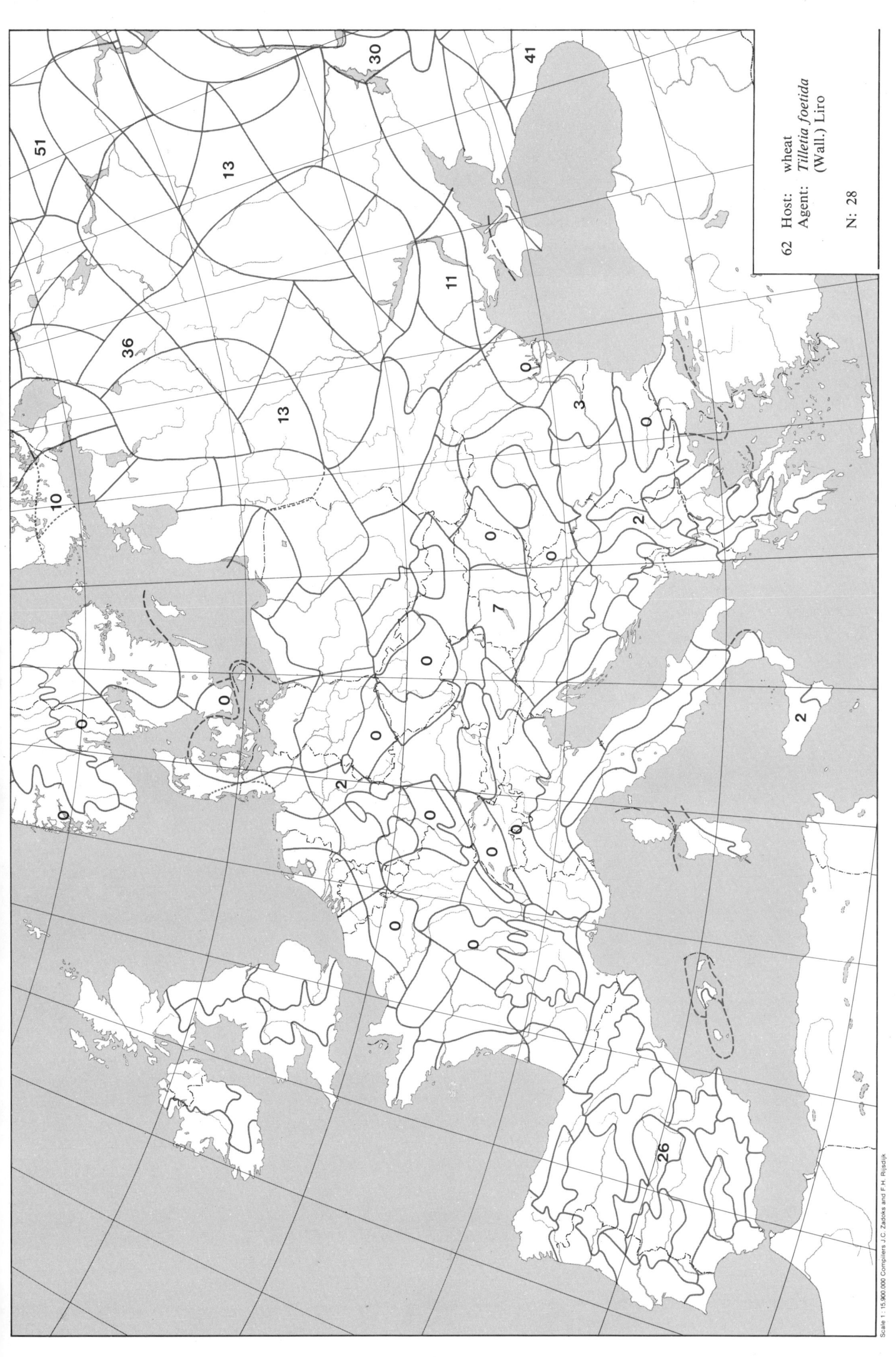

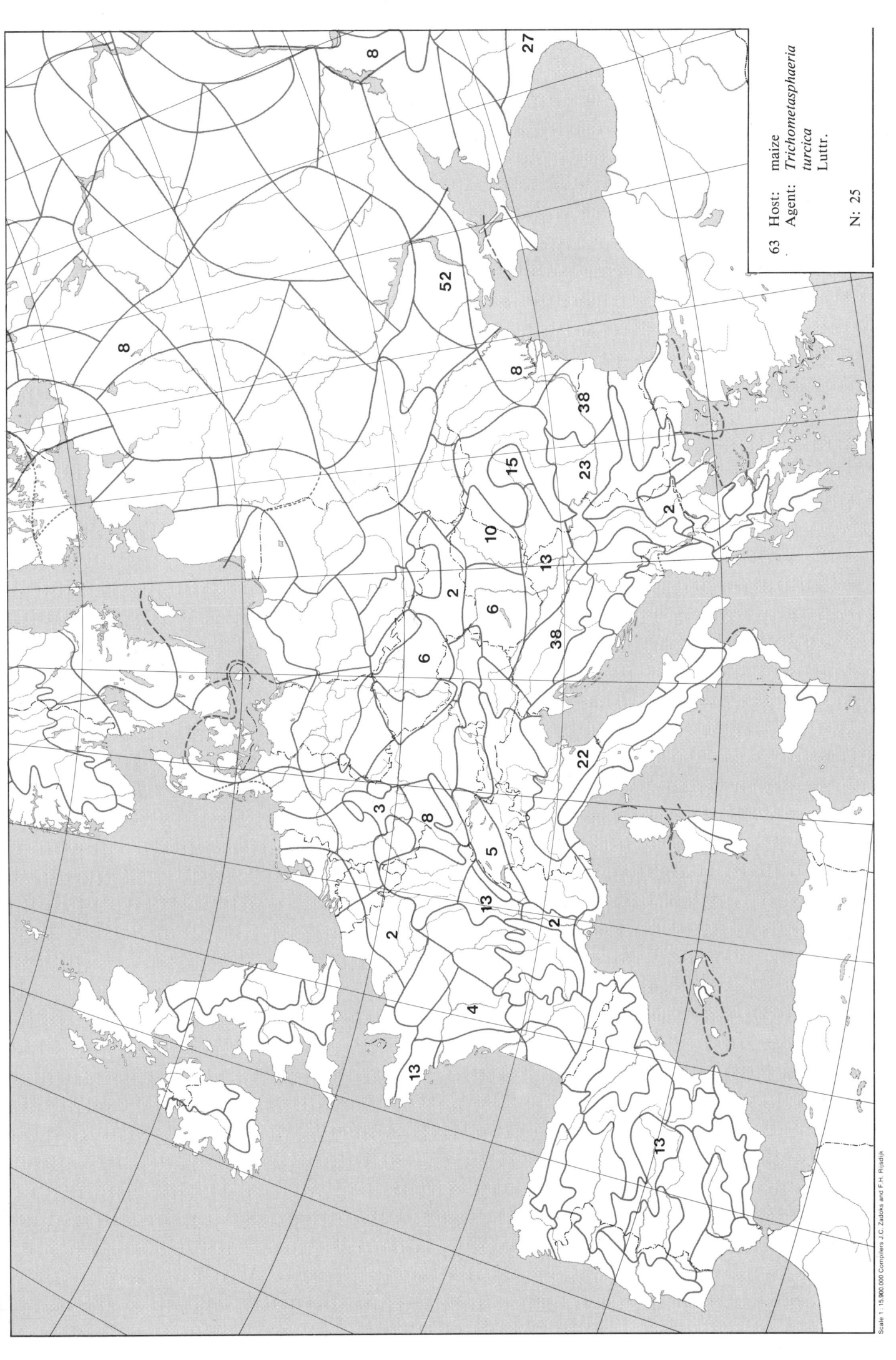

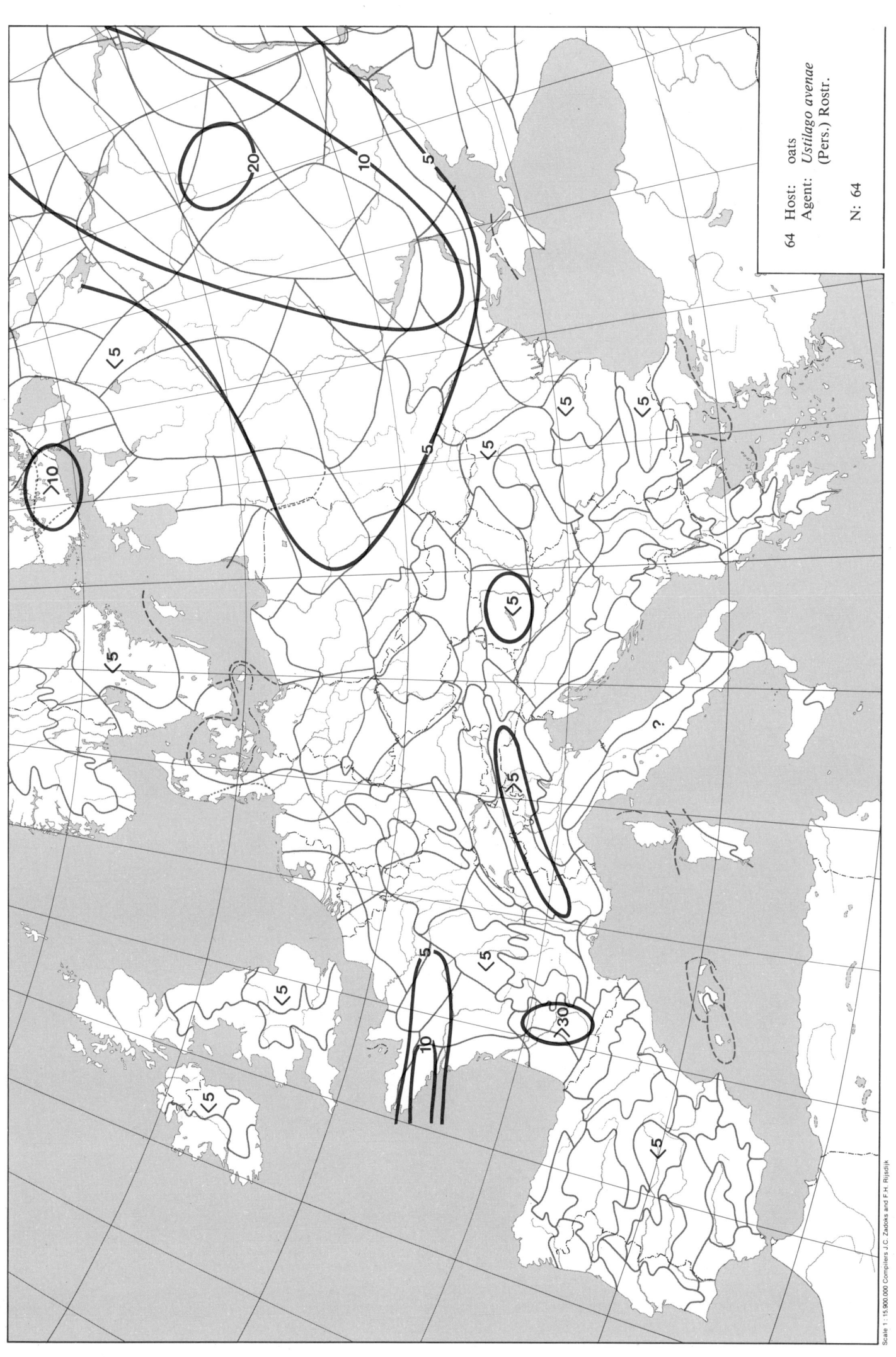

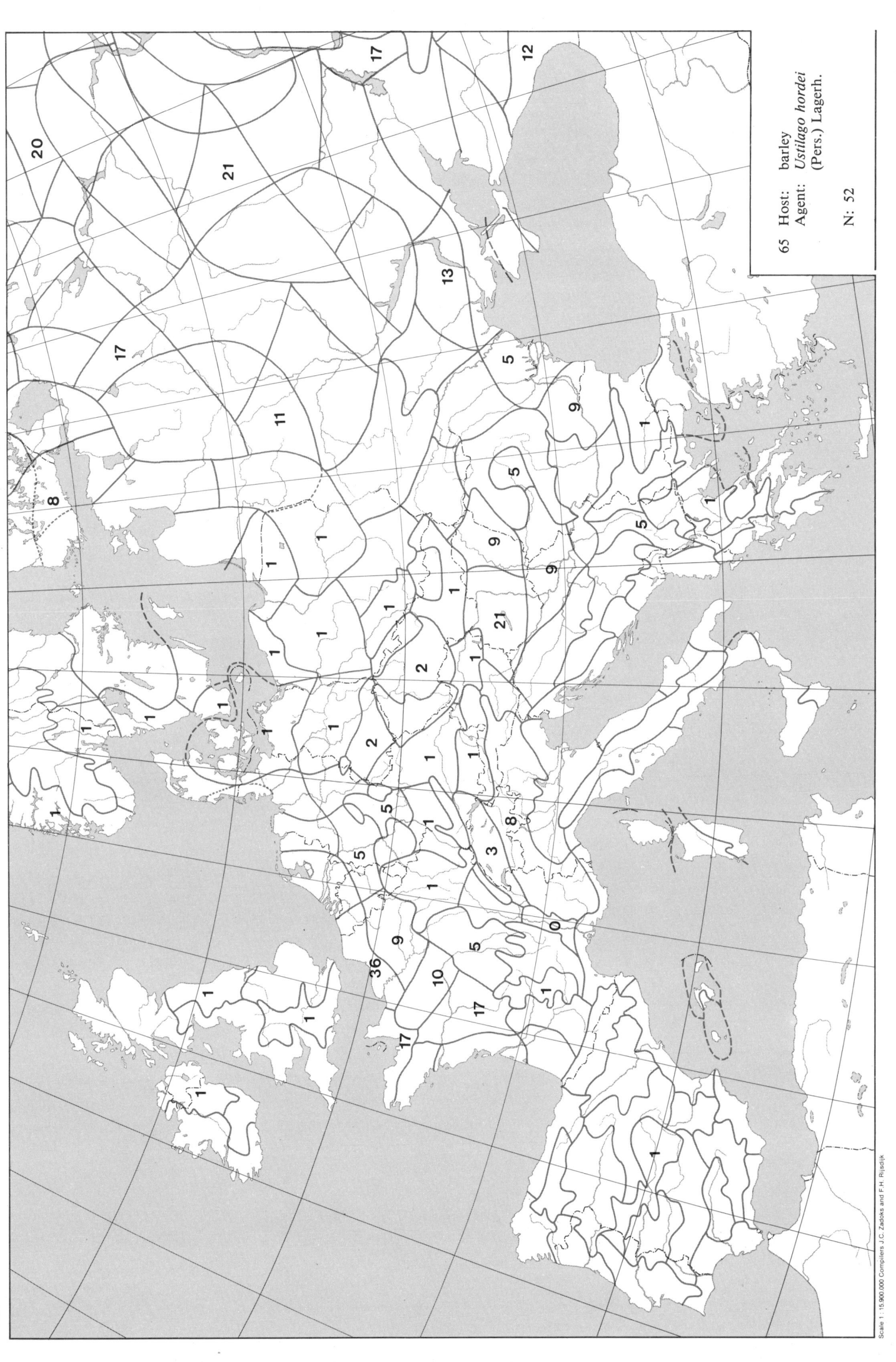

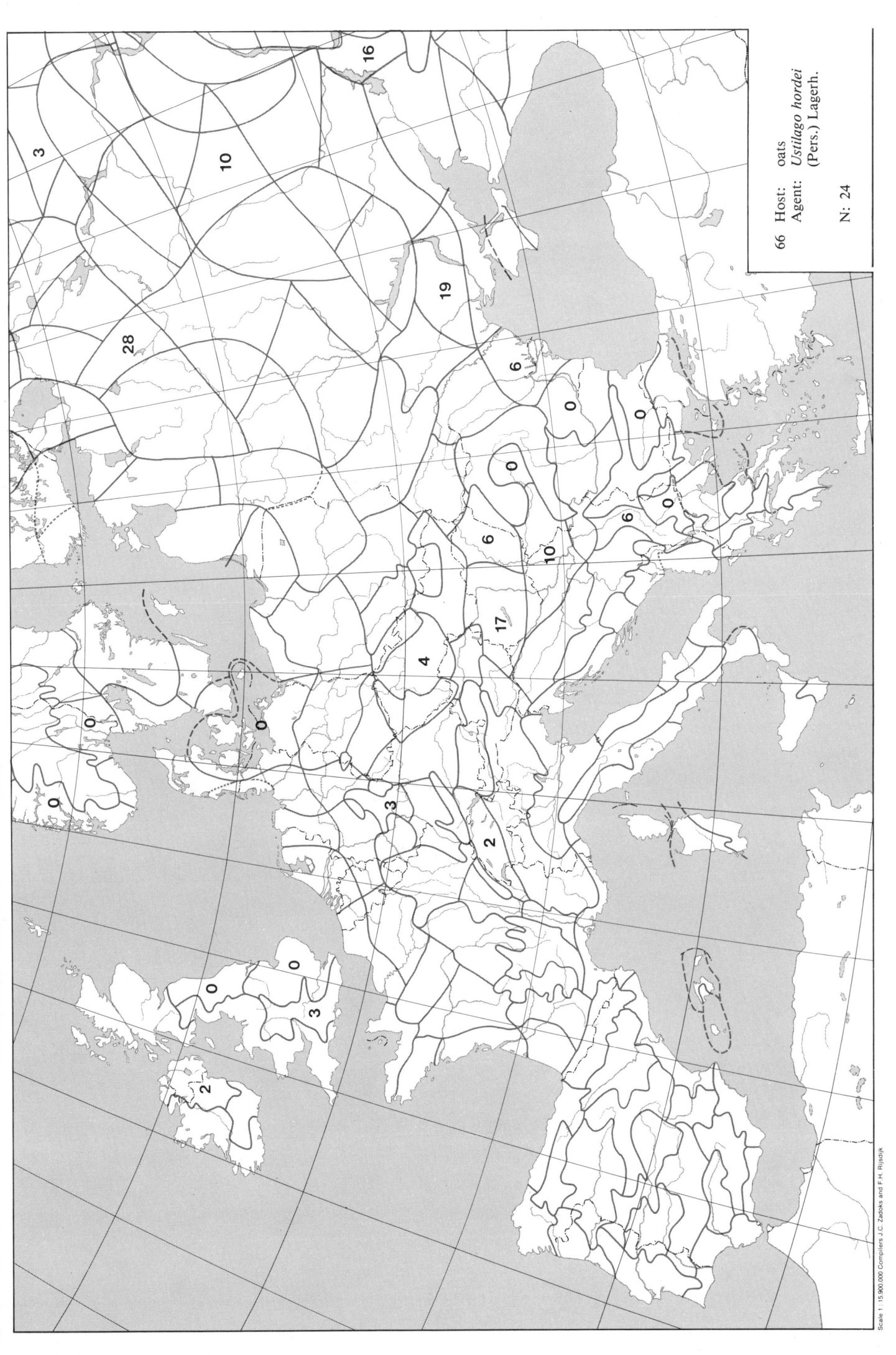

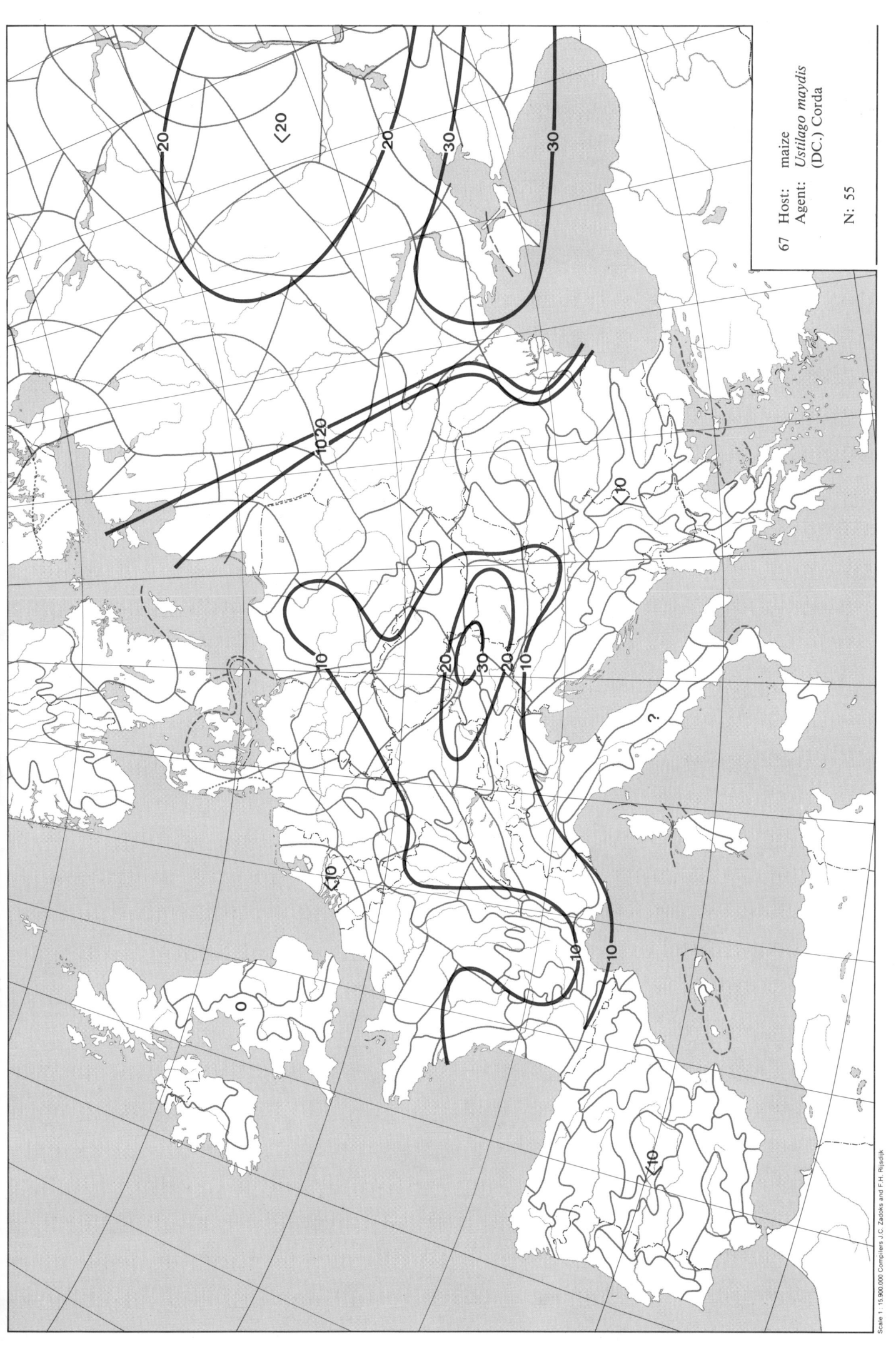

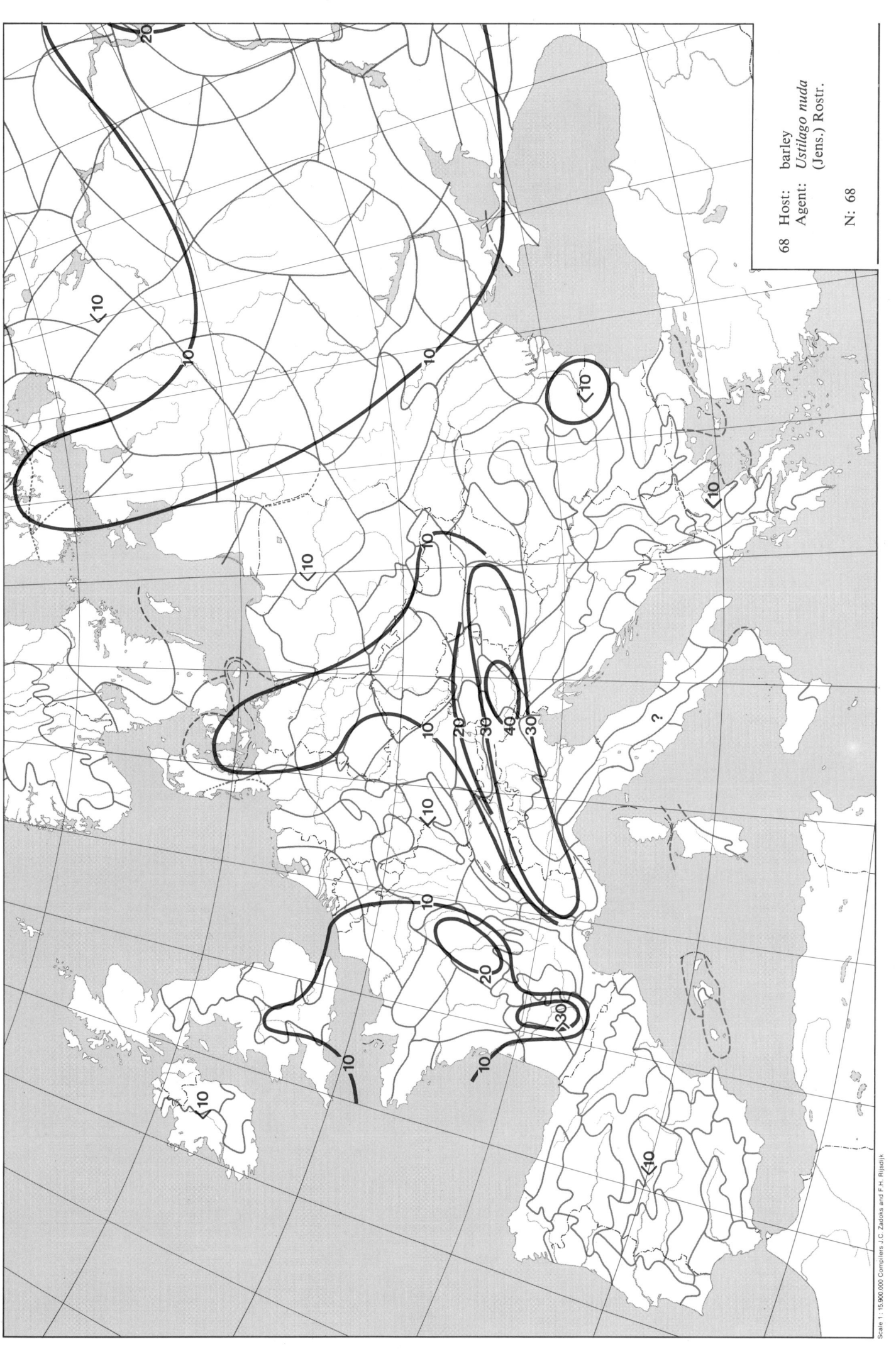

68 Host: barley
    Agent: *Ustilago nuda* (Jens.) Rostr.

    N: 68

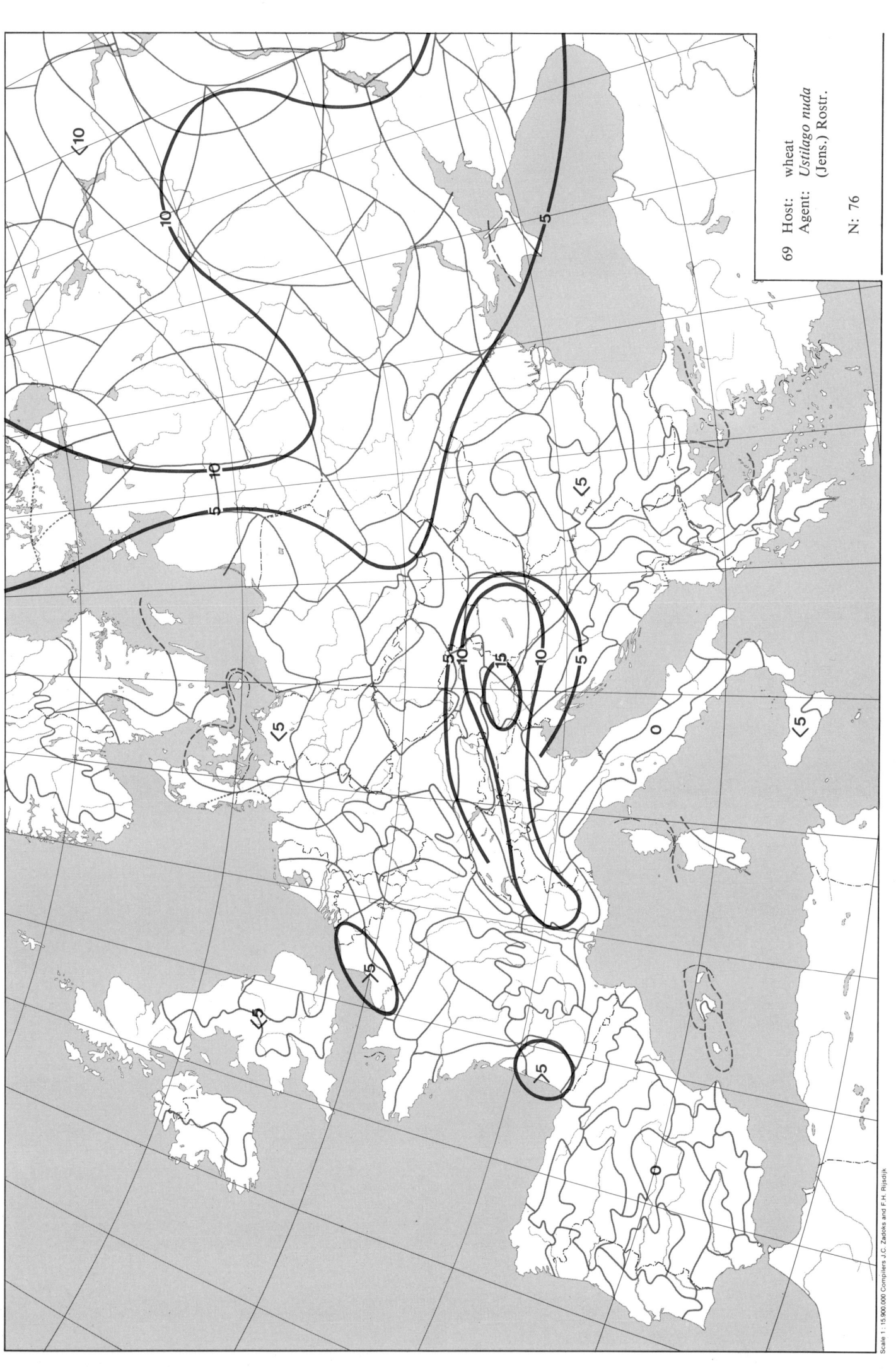

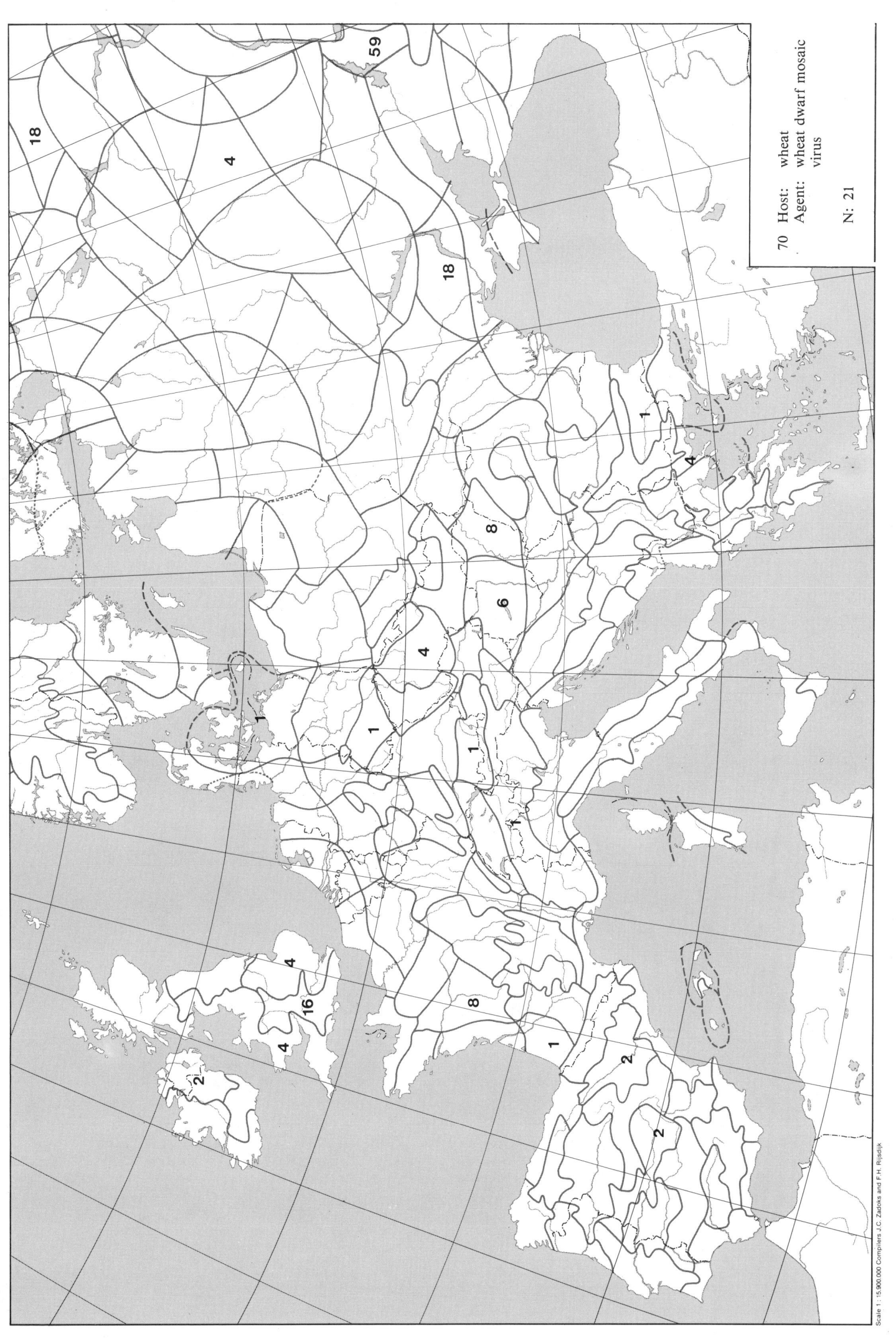

Index of scientific names

(b: barley, m: maize, o: oats, r: rye, w: wheat)

Scientific name	Common name	Map	Crop				
Anguina tritici	(ear) cockles, wheat nematode	2					w
(barley) yellow dwarf virus		3	b				
Cephus pygmaeus	(European wheat stem) sawfly	4					w
Chlorops pumilionis	ribbon-footed cornfly, gout fly	5	b				
Chlorops pumilionis	ribbon-footed cornfly, gout fly	6					w
Cladosporium herbarum	black mould	7					w
Contarinia tritici	lemon wheat blossom midge	8					w
Ditylenchus dipsaci	stem nematode	9				r	
Erysiphe graminis	(powdery) mildew	10	b				
Erysiphe graminis	(powdery) mildew	11					w
Fusarium culmorum	seedling blight, foot rot, root rot, head blight, scab	12					w
Gaeumannomyces graminis	take-all, whiteheads	13	b				
Gaeumannomyces graminis	take-all, whiteheads	14			o		
Gaeumannomyces graminis	take-all, whiteheads	15					w
Gibberella avenacea	seedling blight, brown foot rot, ear blight	16					w
Gibberella zeae	seedling blight, stalk rot, ear rot, scab	17		m			
Gibberella zeae	seedling blight, foot rot, head blight, scab	18					w
Haplodiplosis equestris	saddle gall midge	19	b				
Haplodiplosis equestris	saddle gall midge	20					w
Haplothrips tritici	wheat thrips	21					w
Heterodera avenae	(cereal) cyst nematode	22			o		
Lema spp.	leaf beetle	23			o		
Lema spp.	leaf beetle	24					w
Leptohylemia coarctata	(wheat) bulb fly	25	b				
Leptohylemia coarctata	(wheat) bulb fly	26				r	
Leptohylemia coarctata	(wheat) bulb fly	27					w
Leptosphaeria avenaria	speckled blotch	28			o		
Leptosphaeria nodorum	glume blotch	29					w
Mayetiola destructor	Hessian fly	30				r	
Mayetiola destructor	Hessian fly	31					w
Monographella nivalis	(pink) snow mould, seedling blight, ear blight	32	b				
Monographella nivalis	(pink) snow mould, seedling blight, ear blight	33				r	
Monographella nivalis	(pink) snow mould, seedling blight, ear blight	34					w
Mycosphaerella graminicola	leaf spot, (speckled) leaf blotch	35					w
Oscinella frit	frit fly	36	b				
Oscinella frit	frit fly	37		m			
Oscinella frit	frit fly	38			o		
Ostrinia nubilalis	(European) corn borer	39		m			
Pentatomidae	(shield) bugs (various species, among which 'suni' bug)	40					w
Pentatomidae	(shield) bugs (various species, among which 'suni' bug)	41					w

Scientific name	Common name	Map	Crop			
Pseudocercosporella herpotrichoides	eyespot, stem break	42	b			
Pseudocercosporella herpotrichoides	eyespot, stem break	43				w
Puccinia coronata	crown rust	44		o		
Puccinia graminis	black (stem) rust	45				w
Puccinia hordei	brown (leaf) rust	46	b			
Puccinia recondita	brown (leaf) rust	47			r	
Puccinia recondita	brown (leaf) rust	48				w
Puccinia sorghi	maize (corn) rust	49			m	
Puccinia striiformis	yellow (stripe) rust	50	b			
Puccinia striiformis	yellow (stripe) rust	51				w
Pyrenophora teres	net blotch	52	b			
Rhopalosiphon padi	bird cherry − oat aphid	53		o		
Rhynchosporium secalis	scald, leaf blotch	54	b			
Schizaphis graminum	spring wheat aphid, green bug	55				w
Sitobium avenae	English grain aphid	56		o		
Sitobium avenae	English grain aphid	57				w
Sitodiplosis mosellana	orange wheat blossom midge, red maggot	58				w
Stenothrips graminum	oat thrips	59		o		
Tilletia caries	(common) bunt	60				w
Tilletia controversa	dwarf bunt	61				w
Tilletia foetida	bunt	62				w
Trichometasphaeria turcica	(northern) leaf blight, black ear rot	63			m	
Ustilago avenae	loose smut	64		o		
Ustilago hordei	covered smut	65	b			
Ustilago hordei	covered smut	66		o		
Ustilago maydis	(blister) smut, (common) smut	67			m	
Ustilago nuda	loose smut	68	b			
Ustilago nuda	loose smut	69				w
(wheat) dwarf mosaic virus		70				w

Index of common names

(b: barley, m: maize, o: oats, r: rye, w: wheat)

Common name	Scientific name	Map	Crop			
barley yellow dwarf virus		3	b			
bird cherry – oat aphid	*Rhopalosiphon padi*	53			o	
black ear rot	*Trichometasphaeria turcica*	63		m		
black mould	*Cladosporium herbarum*	7				w
black rust	*Puccinia graminis*	45				w
blister smut	*Ustilago maydis*	67		m		
brown foot rot	*Gibberella avenacea*	16				w
brown rust	*Puccinia hordei*	46	b			
brown rust	*Puccinia recondita*	47				r
brown rust	*Puccinia recondita*	48				w
bugs (various species, among which 'suni' bug)	*Pentatomidae*	40				w
bugs (various species, among which 'suni' bug)	*Pentatomidae*	41				w
bulb fly	*Leptohylemia coarctata*	25	b			
bulb fly	*Leptohylemia coarctata*	26				r
bulb fly	*Leptohylemia coarctata*	27				w
bunt	*Tilletia caries*	60				w
bunt	*Tilletia foedita*	62				w
cereal cyst nematode	*Heterodera avenae*	22			o	
cockles	*Anguina tritici*	2				w
common bunt	*Tilletia caries*	60				w
common smut	*Ustilago maydis*	67		m		
corn borer	*Ostrinia nubilalis*	39		m		
corn rust	*Puccinia sorghi*	49		m		
covered smut	*Ustilago hordei*	65	b			
covered smut	*Ustilago hordei*	66			o	
crown rust	*Puccinia coronata*	44			o	
cyst nematode	*Heterodera avenae*	22			o	
dwarf bunt	*Tilletia controversa*	61				w
dwarf mosaic virus		70				w
ear blight	*Gibberella avenacea*	16				w
ear blight	*Monographella nivalis*	32	b			
ear blight	*Monographella nivalis*	33				r
ear blight	*Monographella nivalis*	34				w
ear cockles	*Anguina tritici*	2				w
ear rot	*Gibberella zeae*	17		m		
English grain aphid	*Sitobium avenae*	56			o	
English grain aphid	*Sitobium avenea*	57				w
European corn borer	*Ostrinia nubilalis*	39		m		
European wheat stem sawfly	*Cephus pygmaeus*	4				w
eyespot	*Pseudocercosporella herpotrichoides*	42	b			

Common name	Scientific name	Map	Crop			
eyespot	*Pseudocercosporella herpotrichoides*	43				w
foot rot	*Fusarium culmorum*	12				w
foot rot	*Gibberella zeae*	18				w
frit fly	*Oscinella frit*	36	b			
frit fly	*Oscinella frit*	37		m		
frit fly	*Oscinella frit*	38			o	
glume blotch	*Leptosphaeria nodorum*	29				w
gout fly	*Chlorops pumilionis*	5	b			
gout fly	*Chlorops pumilionis*	6				w
green bug	*Schizaphis graminum*	55				w
head blight	*Fusarium culmorum*	12				w
head blight	*Gibberella zeae*	18				w
Hessian fly	*Mayetiola destructor*	30			r	
Hessian fly	*Mayetiola destructor*	31				w
leaf beetle	*Lema* spp.	23			o	
leaf beetle	*Lema* spp.	24				w
leaf blight	*Trichometasphaeria turcica*	63		m		
leaf blotch	*Mycosphaerella graminicola*	35				w
leaf blotch	*Rhynchosporium secalis*	54	b			
leaf rust	*Puccinia hordei*	46	b			
leaf rust	*Puccinia recondita*	47			r	
leaf rust	*Puccinia recondita*	48				w
leaf spot	*Mycosphaerella graminicola*	35				w
lemon wheat blossom midge	*Contarinia tritici*	8				w
loose smut	*Ustilago avenae*	64			o	
loose smut	*Ustilago nuda*	68	b			
loose smut	*Ustilago nuda*	69				w
maize rust	*Puccinia sorghi*	49		m		
mildew	*Erysiphe graminis*	10	b			
mildew	*Erysiphe graminis*	11				w
net blotch	*Pyrenophora teres*	52	b			
northern leaf blight	*Trichometasphaeria turcica*	63		m		
oat thrips	*Stenothrips graminum*	59			o	
orange wheat blossom midge	*Sitodiplosis mosellana*	58				w
pink snow mould	*Monographella nivalis*	32	b			
pink snow mould	*Monographella nivalis*	33			r	
pink snow mould	*Monographella nivalis*	34				w
powdery mildew	*Erysiphe graminis*	10	b			
powdery mildew	*Erysiphe graminis*	11				w
red maggot	*Sitodiplosis mosellana*	58				w
ribbon-footed corn fly	*Chlorops pumilionis*	5	b			
ribbon-footed corn fly	*Chlorops pumilionis*	6				w
root rot	*Fusarium culmorum*	12				w
saddle gall midge	*Haplodiplosis equestris*	19	b			
saddle gall midge	*Haplodiplosis equestris*	20				w
sawfly	*Cephus pygmaeus*	4				w
scab	*Fusarium culmorum*	12				w
scab	*Gibberella zeae*	17		m		
scab	*Gibberella zeae*	18				w
scald	*Rhynchosporium secalis*	54	b			
seedling blight	*Fusarium culmorum*	12				w
seedling blight	*Gibberella avenacea*	16				w
seedling blight	*Gibberella zeae*	17		m		
seedling blight	*Gibberella zeae*	18				w
seedling blight	*Monographella nivalis*	32	b			
seedling blight	*Monographella nivalis*	33			r	

Common name	Scientific name	Map	Crop			
seedling blight	*Monographella nivalis*	34				w
shield bugs (various species, among which 'suni' bug)	*Pentatomidae*	40				w
shield bugs (various species, among which 'suni' bug)	*Pentatomidae*	41				w
smut	*Ustilago maydis*	67	m			
snow mould	*Monographella nivalis*	32	b			
snow mould	*Monographella nivalis*	33			r	
snow mould	*Monographella nivalis*	34				w
speckled blotch	*Leptosphaeria avenaria*	28		o		
speckled leaf blotch	*Mycosphaerella graminicola*	35				w
spring wheat aphid	*Schizaphis graminum*	55				w
stalk rot	*Gibberella zeae*	17	m			
stem break	*Pseudocercosporella herpotrichoides*	42	b			
stem break	*Pseudocercosporella herpotrichoides*	43				w
stem nematode	*Ditylenchus dipsaci*	9			r	
stem rust	*Puccinia graminis*	45				w
stripe rust	*Puccinia striiformis*	50	b			
stripe rust	*Puccinia striiformis*	51				w
take-all	*Gaeumannomyces graminis*	13	b			
take-all	*Gaeumannomyces graminis*	14		o		
take-all	*Gaeumannomyces graminis*	15				w
wheat bulb fly	*Leptohylemia coarctata*	25	b			
wheat bulb fly	*Leptohylemia coarctata*	26			r	
wheat bulb fly	*Leptohylemia coarctata*	27				w
(wheat) dwarf mosaic virus		70				w
wheat midge	*Sitodiplosis mosellana*	58				w
wheat nematode	*Anguina tritici*	2				w
wheat thrips	*Haplothrips tritici*	21				w
whiteheads	*Gaeumannomyces graminis*	13	b			
whiteheads	*Gaeumannomyces graminis*	14		o		
whiteheads	*Gaeumannomyces graminis*	15				w
yellow dwarf virus		3	b			
yellow rust	*Puccinia striiformis*	50	b			
yellow rust	*Puccinia striiformis*	51				w